WHAT SHIP
WHERE BOU

Norfolk Museums Service (Time and Tide Museum)

A HISTORY OF VISUAL
COMMUNICATION AT SEA

DAVID CRADDOCK

Seaforth
PUBLISHING

First published in Great Britain in 2021 by
Seaforth Publishing,
A division of Pen & Sword Books Ltd,
47 Church Street,
Barnsley S70 2AS
www.seaforthpublishing.com

British Library Cataloguing in Publication Data
A catalogue record for this book is available from the British Library

ISBN 978 1 5267 8482 7 (PAPERBACK)
ISBN 978 1 5267 8483 4 (EPUB)
ISBN 978 1 5267 8484 1 (KINDLE)

Pen & Sword Books Limited incorporates the imprints of Atlas, Archaeology, Aviation, Discovery,
Family History, Fiction, History, Maritime, Military, Military Classics, Politics, Select, Transport, True
Crime, Air World, Frontline Publishing, Leo Cooper, Remember When, Seaforth Publishing, The
Praetorian Press, Wharncliffe Local History, Wharncliffe Transport,
Wharncliffe True Crime and White Owl

Typeset and designed by David Craddock.
Printed and bound in India by Replika Press Pvt Ltd

Acknowledgements
Many people have contributed to this book, not least those who pioneered the signal flags,
semaphore arms and flashing lights that are its subject, for which grateful thanks are due. The
author owes a special debt of gratitude to Dr Jane Harrold and Dr Richard Porter and fellow
Trustees of the Britannia Museum at the Royal Naval College, Dartmouth for access to the
museum's collection; to Dr Michael Duffy, former Director of the Centre for Maritime Historical
Studies at the University of Exeter, for his time in reading and advising on sections of the book; to
the librarians at The Caird Library, National Maritime Museum and the ever helpful staff at The
National Archives in Kew. Thanks also to former Royal Navy 'bunting' David Morris; to Malcolm
Dobson, the name of whose company, Francis Searchlights, will be found on marine signal lamps
worldwide and to Julian Mannering at Seaforth Publishing for his encouragement and advice.

While every effort has been made to trace and credit copyright holders of images used, some
cannot be attributed with certainty, nevertheless the use of all are gratefully acknowledged.

A note on spelling
the term 'pennant' occurs in many places describing an elongated flag tapering towards the fly.
Texts up to the mid-20th century tend to use the word 'pendant' to describe the same thing
though it is usually pronounced 'pennant'. For the sake of consistency, I have used the more
frequently found modern spelling throughout.

Contents

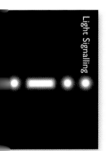
Flag Signalling

Semaphore

Light Signalling

Introduction

The question in the title of this book will be familiar to many who have exchanged messages by signal lamp on shipping routes around the world. The starting point for the book was the author's own instruction in marine signalling under Chief Petty Officer 'Charlie' Sewell – a boy signaller aboard HMS *Neptune* at the Battle of Jutland – at Pangbourne Nautical College, prior to joining P&O as a cadet. A later career as a graphic and exhibition designer, with a particular focus on our maritime heritage, has both sustained the interest and informed the research on which this book is based.

Flotilla Leader HMS *Blencathra*, a 'Hunt' class destroyer flying her pennant number, photographed without armaments late 1940.

The subject of visual communications at sea is huge, spanning, at least in written record, two and a half millennia. It is a story dominated by the development of flag signalling at sea with the early codification of signals inextricably linked with fleet manoeuvres and war fighting under sail. This is as it should be for the imperative of commanders to communicate unambiguously with their fleets has been the driver of innovation and experiment explored here. Semaphore at sea came later and reliable signalling by light not until the end of the 19th century. All three methods are still in use and have left a legacy that has become part of our visual and cultural heritage.

From the turn of the 19th century the growing importance of the mercantile market gave rise to dozens of competing codes for communication at sea and between ships and signal stations ashore, employing flags, lights, shapes and sometimes combinations of all three. Few found wide acceptance and by mid-century the British Board of Trade, in publishing the first edition of *The Commercial Code* in 1857, paved the way for what would become the International Code of Signals still in use today. While not wanting to give priority here to one means of signalling over another, for each serve distinct purposes, the long history of flag signalling at sea inevitably dominate. Though reference is made to practice among other maritime nations, this book primarily reflects signalling practice in the Royal Navy and the British merchant service.

A new study of the subject must acknowledge earlier scholarship and here it is to W G Perrin and his 1922 work *British Flags, their Early History and their Development at Sea* that the greatest debt is owed, as it is to his 1908 study of Nelson's famous signal at Trafalgar. Another landmark work is *Signal, A History of Signalling in the Royal Navy* by Captain Barrie Kent. On a lighter note, Captain Jack Broome's 1956 *Make a Signal* and the later *Make Another*

Signal celebrate the dry wit that often characterised naval signal exchanges. All are cited either in endnotes or in the bibliography. Also acknowledged at the end of the book are a number of copiously illustrated on-line resources that will fill in the gaps in this account and satisfy the most curious vexillologist.

All signalling methods had their limitations which gave rise to unintended consequences, some of which are explored here. Most signal exchanges followed the prescriptions of the *Signal Book*, but not all. In both Services there are anecdotal accounts of off-the-record exchanges, sometimes cryptic, nearly always good humored. A 1941 exchange of signals between a flotilla leader, Captain (D) Philip Ruck-Keene aboard HMS *Blencathra* (see left), and his old friend Rhoderick 'Wee Mac' McGrigor, Flag Captain aboard the battlecruiser HMS *Renown* (right) was typical of the signal sparring that relieved the tedium of routine patrols. Ruck-Keene was a man of considerable stature and his plain-language morse signal: '**What a gigantic contraption for such a very small driver**' brought an immediate riposte from the rather shorter McGrigor: '**While big apes cling to smaller branches**', a reference to the Royal Navy's vital destroyer branch. McGrigor went on to become First Sea Lord.[1]

The battlecruiser HMS *Renown* in 1939.

Radio and satellite communications have reduced the dependency on visual signalling but they are not always secure and, at close quarters, flag hoists, semaphore and light still have a role. Experimentation in the US Navy with text messaging using high-speed LED signal lamps is not something that Samuel Morse could have anticipated but, just as demands for more reliable communication in the eighteenth century drove flag design and coding methods, so do the same pressures drive experimentation in secure communication now.

As this book will show, the history of visual signalling, particularly with flags, has been a long process of iteration, with more than ten different sets of numeral flags in use between 1778 and 1900 alone. Though perhaps now more familiar on tea towels, mugs and cushion covers, many code flags still in use internationally today have their origins in the last decades of the eighteenth century. Some of the signals they have conveyed are embedded equally in our maritime heritage and popular culture, a rich legacy celebrated in these pages.

Reinier Nooms, *The Battle of Leghorn 1653* (detail). Rijksmuseum, Amsterdam

66 Now, my dear Kempy, do, for God's sake,
oblige me by throwing your signals overboard and
make that which we all understand:
'BRING THE ENEMY TO CLOSE ACTION!' 99

Admiral Sir Charles Geary to his Flag Captain Richard Kempenfelt
aboard HMS *Victory*, blockading Brest in July 1780.

Flag Signalling Introduction

Of all the mediums for visual signalling at sea: flags, semaphore and by light, signalling by flag was developed entirely by and for seafarers. The timeline on the pages that follow traces the use of flags at sea from the basic requirement of the admiral to call a council of his captains to the codification of complex manoeuvres embodied in the work of Admirals Hawke, Howe, Kempenfelt and Rodney. Many names will be familiar, others less so; for this is a story that, while focusing on the development of flag design and signalling practice in the Royal Navy, extends beyond our shores to acknowledge the advocacy and tactical thinking of allies and former adversaries.

Early records of flag signals reflect the more highly developed naval warfare between Mediterranean states where galleys and combined fleets of sail and oared vessels both enabled and required the coordination of complex manoeuvres. In Nathaniel Boteler's *Sea Dialogues* (1627-34) we get the first proposals in Britain for specifically coloured signal flags, but it was the pressure of the Dutch wars later in

the seventeenth century that saw the issue of the 'Commonwealth Code' and 'Fighting Instructions' under the direction of General-at-Sea Sir Robert Blake. The first systematic tabulation of signals was included in the 1673 *Articles of Sailing and Fighting Instructions* which formed the basis upon which signalling at sea evolved over the following century. Code flags included in the 1673 Articles are shown on pages 18 and 19, as are many of the variations of colour and design that found their way into the flag lockers and signal books of HM ships.

The second half of the 18th century, often marked by conflict at sea, saw a gradual move from the haphazard introduction of new signals by individual commanders-in-chief to a settled consensus captured in the final iteration of Admiral Lord Howe's 1799 *Signal Book for the Ships of War*. But it was not without some resistance as the good-humoured exchange between Admiral Geary and his Flag Captain at the foot of the previous page shows.

From the turn of the 19th century and the introduction of Popham's *Marine Vocabulary,* the pace quickened. By mid-century, several flag systems competed for the lucrative commercial market which naval signalmen were also required to learn. At the same time the sea-going adaptations of the new land telegraphs, described in the later sections of this book, added to their workload. This is therefore as much the signalman's story as that of the fleet tactician and colour theorists who compiled the signal book.

The illustration below is one of 52 engravings commissioned from marine artist T L Hornbrook for an 1834 English translation by Captain J D Boswall RN of Père Paul Hoste's Treatise on Naval Tactics. Hoste's original 1697 *Traité des Évolutions Navales* was first published in English in 1763 and was almost certainly influential in British tactical thinking and the signals by which fleet evolutions were conducted. See also page 41.

Google Books

Timeline: Councils, codes and competition

c.450 BC　Historians **Herodotus** and **Thucydides** both refer to *semeion* (σημήιον), a sign or signal in connection with sea battles in the Persian and Peloponnesian wars. One such sign is recorded as being made more conspicuous by the inclusion of a 'military cloak of a bright colour'. This may be the origin of a sign known as *Phoinikis* (Φοινικίς), made with a purple-red dye extracted from sea snails in the eastern Mediterranean; hence Phoenician.

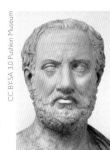

Greek historian of the Peleponnesian War Thucydides, a plaster cast from a Roman copy of the original from the early 4th century BC.

c.900 AD　Earliest known code of naval signals set out in the nineteenth chapter of Byzantine Emperor **Leo VI's** *Tactica* which calls for "...either a banner or a streamer... in some conspicuous position... that you may be able thereby to make known what requires to be done."

c.1338　The *Black Book of the Admiralty* includes two specific signals, most likely in the form of a banner: one for the admiral to call a council of his captains and a second for signalling the presence of an enemy.

Mosaic of Emperor Leo VI, also known as Leo the Wise in the Hagia Sophia, Istanbul.

1366　**Amadeo de Savoy** issues set of instructions for warfare in galleys with flag signals to address specific ships, to report sighting the enemy, for summoning help and for recognition.

1420　**Giovani Mocenigo**, Doge of Venice, issues orders for the Venetian navy, at war with the Ottomans, which contain details of flag shape and position.

1430　Castillian Admiral **Fadrique Henriques** issues similar instructions to his fleet which include signals for summoning assistance.

Bronze relief of Giovani Mocenigo, Doge of Venice.

1515　French Admiral **Antoine de Conflans** issues *Ordonnances et signes pour naviguer jour et nuyt en une armée royale* – a comprehensive set of day and night signals, including flashing lights, for a fleet combining sailing ships and galleys but there is no evidence they were actually used.

1517	Instructions issued by **Philippe de Cleves** and used by Emperor Charles V in voyage from Flanders to Spain in 1517 and incorporated into **Jehan Bytharne's** *Livre de Guerre tant par la mer que par terre* in 1543.

c.1530 · Henry VIII's Lord Chancellor **Thomas Audley** issues Orders to be used in the 'King's Majesties Navy by the Sea' which begins with the instruction for summoning a council:

> "Whensoever, and at all tymes the Admyrall doth shote a pece of Ordinance, and set up his Banner of Councell on Starrborde bottocke [quarter] of his Shippe, everie shipps capten shall with spede go aboard the Admyrall to know his will"

1792 engraving of Thomas Audley by Peltro William Tomkins after a painting by Hans Holbein the Elder.

1627-1634 · **Nathaniel Boteler**, who sailed on the disastrous 1625 Cadiz Expedition and was a member of the Virginia Company, publishes *Six Dialogues about Sea Service between an High Admiral and a Captain at Sea*, the sixth of which includes signals 'for ordering of the fleet'. His proposals include use of tops'ls for distance signalling and special flags for summoning a council, making landfall or sighting the enemy.

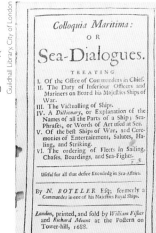

Title page from Nathaniel Boteler's *Dialogues* re-published as *Colloquia Maritima or Sea-Dialogues* in 1688.

1647 · **The Right Honourable Committee of Lords and Commons for the Admiralty and the Cinque Ports** issues new Instructions for Sailing which bring the Navy's signalling practice more into line with that of the French and Spanish navies with the introduction of the first codes.

1653 · 'Commonwealth Code' issued by **The Right Honourable the Generals and Admirals of the Fleet** under the guidance of General-at-Sea **Sir Robert Blake** whose *Fighting Instructions* are the first to adopt the Line of Battle.

1665 · Commonwealth Code expanded under **Duke of York**, later King James II.

1673 · New instructions issued: *Articles of Sailing and Fighting Instructions* include the first tabulated list of signals with instruction from where they were to be shown and their meaning.

General-at-Sea Sir Robert Blake, often referred to as 'The Father of the Navy' painted in 1829 by Henry Perronet Briggs.

| 1691 | Instructions of 1673 revised and updated by **Admiral Edward Russell** and again in 1703 by **Admiral Sir George Rooke**. |

1693   French Admiral **de Tourville** publishes illustrated signal book with each signal hoist shown in its correct place.

1714   *Sailing and Fighting Instructions or Signals as they are Observed in the Royal Navy of Great Britain* is the first signal book to be printed unofficially in pocket-book form by **Jonathan Greenwood** as a private venture. It followed a similar pattern to de Tourville's book and is dedicated to Admiral Russell.

Admiral Anne
Hilarion de Tourville
by unknown artist.

1738   Captain **Mahé de la Bourdonnais** devises the first numerical code flag system for use by ships of the French East India Company. It was never published, but described by **Bourdé de Villehuet** in his 1769 classic on tactics *Le Manoeuvrier*.

1746   **John Millan** privately publishes an illustrated pocket *Signal Book for the Royal Navy*, 2s. 6d. plain and 4s. coloured, with different types of signal listed under the headings 'Sailing', 'Fighting' and 'Additional Instructions'.

Title page from
John Millan's
pocket guide,
1746.

1756   New manuscript signal book issued, probably by **Admiral Sir Edward Hawke**, and used at the Battle of Quiberon Bay in November 1759 in which he boldly inflicted a heavy defeat on the French fleet. He introduces a number of new flags including the first chequered flag.

1762   The Admiralty issues illustrated manuscript signal book, adding new flags, based on Hawke's earlier signal book under the title *General Printed and Additional Signals delivered out by Sir Edward Hawke*.

1763   **Sébastian Bigot de Morogues** publishes his *Tactique Navale ou Traité des Evolutions et des Signaux*, a detail study of sea warfare, fleet manoeuvres and signalling which is widely circulated with translations into English and Dutch.

Page from Bigot de Morogue's 1763 *Tactique Navale*. Fleet evolutions were numbered and each allocated a numerical signal.

1776	**Vice Admiral Lord Howe**, as C-in-C North American Station, issues his first signal book from his flagship HMS *Eagle* at Sandy Hook, assigning meanings – all concerning manoeuvring of ships – to single flags flown from specific positions. This was revised in 1778 to a numerical system divided into signals from the admiral and those from private i.e. non-flag ships.
1777	**Admiral Sir Charles Knowles**, then a lieutenant, publishes his *Set of Signals for a Fleet on a Plan Entirely New*. It includes a code based on a 'chessboard' matrix, first proposed by Mahé de la Bourdonnais in 1738 and later published in 1769 by Bourdé de la Villehuet.

1780 **Rear Admiral Richard Kempenfelt** as Chief of Staff of the Channel Fleet under Admiral Sir Charles Hardy issues a signal code based on Lord Howe's numerical signals in specific positions but combining signals for flagships and private ships in a single list. It is re-issued the following year with some changes to make flags more distinct, including the first triangular flags.

Royal Museums Greenwich (BHC 2818)

Portrait (detail) of Rear Admiral Richard Kempenfelt painted by Tilly Kettle in 1782, the year in which Kempenfelt was lost in an accident aboard HMS *Royal George*.

1782 **Admiral George Rodney** adds further flags to Howe's 1778 code book and uses it to good effect at The Battle of the Saintes where a fortuitous wind shift allowed his fleet to break the French line in two places. In the same year Admiral Howe issues a revised code on taking command of the Channel Fleet. One of his captains, John Jervis, later to become Earl St Vincent after his victory at Cape St Vincent in 1797, wrote in September 1782 of the way in which fleet manoeuvres were '...materially aided by the admirable code of day signals which his Lordship had then lately introduced'.

Royal Museums Greenwich (BHC 0701)

Battle of the Saintes painted late 18th century by Thomas Luny showing the French flagship *Ville de Paris* (left foreground), under heavy fire from Rodney's flagship HMS *Formidable*.

1782 Admiral of the Red **Robert Digby**, in command of the North American Station, issued his own signal code, modelled on Admiral Kempenfelt's code. Kempenfelt himself was drowned in a tragic accident aboard his flagship HMS *Royal George* in Portsmouth Harbour in August of that year.

1790
John McArthur, purser aboard the *St Margarita* with Admiral Digby on the North American Station, devised a signal code which included a security device to enable continuous change. Although championed by Vice Admiral Samuel Hood, then First Naval Lord, it was not taken further in deference to Lord Howe whose second signal book was issued in the same year. McArthur later re-organised and simplified Howe's signal book '...engrafting in the body and instructions many new ideas and instructions of his own', which was approved by Howe and issued with some revisions as the *Signal Book for the Ships of War* in 1793. It was this code that was used successfully at the Battle of 1st June in 1794 and the Battles of Cape St Vincent in 1797 and The Nile in 1798. Its publication brought to an end the haphazard system by which individual admirals devised and issued their own codes during the second half of the 18th century.

Admiral of the Fleet Lord Richard Howe by John Singleton Copely, 1794.

Page from 1799 *Signal Book for the Ships of War* published by Henry Edles in the East India Company Presidency of Madras, modern-day Chennai. Britain and France were then at war with their possessions in India heavily contested.

1799
The Admiralty authorises a new and enlarged edition of the *Signal Book for the Ships of War* based on Howe's issues of 1790 and 1793, which individual C-in-Cs were required to supplement with two pennant hoists to identify individual ships within their squadron.

1800
Sir Home Riggs Popham proposes a vocabulary of flag signals: *Telegraphic Signals or Marine Vocabulary* when serving aboard HMS *Romney* in the North Sea squadron under Vice Admiral Sir Archibald Dickson. Note that 'Telegraphic' in the title literally means distance or far-off writing and its use pre-dates the electric telegraph by some 30 years. The first version, printed in 1801, contained approximately 1,000 words and phrases selected for naval purposes only and utilised the numeral flags already coded in the Admiralty's 1799 signal book.

1806 engraving of Sir Home Riggs Popham with title page of his *Marine Vocabulary* published in 1803.

803
With success of his experiments with the North Sea Squadron, Popham added two more sections to his *Marine Vocabulary* containing sentences '...most applicable to military or general conversation.' This marked

a significant shift from a system of purely numeric codes with specific meanings to one more closely related to spoken language.

1805   Popham's 1803 code book, with substituted numeral flags, after the original signal code had been captured from the schooner HMS *Redbridge* by a French frigate off Toulon in the previous year, was used to make Admiral Lord Nelson's famous signal before battle was joined at Trafalgar.

1808   The Admiralty re-issues revised version of 1799 signal book but retaining the same (revised) numeral flags as the 1803 issue. In the same year **Colonel John McDonald** publishes his *Treatise on Telegraphic Communication Naval, Military and Political,* a critique of current systems in use by the Admiralty and proposing his own numerical dictionary for encoding messages either by flag or semaphore.

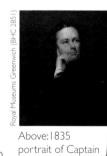

A TREATISE
ON
TELEGRAPHIC COMMUNICATION,
Naval, Military, and Political:
IN WHICH
THE KNOWN DEFECTS OF THE PRESENT SYSTEM OF TELEGRAPHIC PRACTICE BY SEA AND LAND ARE OBVIATED
BY THE INTRODUCTION OF A
Numerical Portable Dictionary,
Calculated, when applied to various described Telegraphs, and to the Naval Flag System, to be an accurate Medium of carrying on distant Conversation, without any Liability to Confusion, Error, or Mistake :
WITH
SOME CONSIDERATIONS ON THE PRESENT STATE OF THE MARINE CODE, AND OF NAVAL SIGNALS.
ILLUSTRATED BY LINEAR PLATES
Connected with the Detail of
THE NEW TELEGRAPHIC SYSTEM ;
Substituting, on very simple Principles,
A SPEAKING, IN LIEU OF A SPELLING POWER,
IN
Different Day and Night Maritime, Civil and Military
Telegraphs.
Dedicated, by Permission, to the Right Honourable
LORD HAWKESBURY.
By JOHN MACDONALD, Esq. F.R.S. F.A'S.
LATE LIEUT. COL. AND ENGINEER, &c. &c.
Frustra fit per plura, quod fieri potest per pauciora.
LONDON ;
PRINTED FOR T. EGERTON,
MILITARY LIBRARY, NEAR WHITEHALL.
1808.

Title page of Colonel John McDonald's 1808 *Treatise.*

1812/13   Popham fully revises his code with 223,675 possible combinations in two, three or four flag hoists. The word 'confides' which had to be substituted with 'expects' at Trafalgar is now included as a three-flag hoist, whereas the latter required four flags.

1816   Admiralty issues official *Vocabulary Signalling Book* based on Popham's work, using 14 letter flags and nine numerals. It includes instructions for Popham's sea-going semaphore (see p. 64).

1817   **Captain Frederick Marryat RN** publishes the first edition of his *Code of Signals for the Merchant Service.* Using just 16 flags it was divided into six sections: English Warships; Foreign Warships; English Merchant Ships (Lloyd's List); Lighthouses, Ports and hazards; Selection of Sentences; Vocabulary. Marryat's code was published in several editions in his lifetime and subsequently by **J M Richardson** in several further editions after his death in 1848.

Royal Museums Greenwich (BHC 2851)

Above: 1835 portrait of Captain Frederick Marryat by E Dixon. Right: Cover of 1861 edition of Marryat's code published by J M Richardson.

1827   The Admiralty publishes a revised edition of 1816 signal book which includes an updated semaphore code. Further editions are published in 1834 and 1859, with some meanings changed to take account of the transition from sail to steam and it is reprinted with further revisions in 1868, 1879 and 1882.

1845   **Henry J Rogers** of Baltimore, Maryland, develops *The Telegraphic Dictionary and Seaman's Signal Book* to encode messages for signalling by flag and shapes or transmission by electric telegraph. It carries a testimonial from Samuel Morse congratulating him on his achievement.

1855   **Capitaine Charles de Reynold de Chauvancy** publishes his *Télégraphie Nautique* with the French government insisting, unsuccessfully, on its use in place of Marryat. Nevertheless, it is translated into six languages including English and published with the approval of the Admiralty in London in the same year as *Reynold's Code: Polyglot Nautical Telegraph for the use of Men of War and Merchant Vessels.*

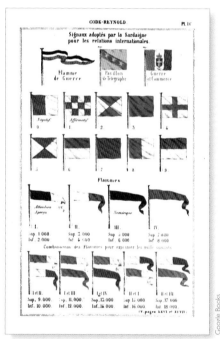

Reynold's 1855 code was widely translated. Shown here is a page from the Italian translation.

1857   British Board of Trade publish *The Commercial Code*, the first edition of what will eventually become the International Code of Signals. It uses only 18 alphabetical flags, omitting vowels to avoid spelling out inadvertent insults, as well as the last three letters of the alphabet. There are no numeral flags. The Admiralty adopted the same letters but used different flags.

887   The *Commercial Code* revised by the Board of Trade and used for basis of discussion among maritime powers at the 1889 Washington International Maritime Conference. Several amendments were made and a new *Commercial Code*, now with the full alphabet, but still omitting numeral flags, was published in 1897 and officially adopted on 1st January 1902.

889   The Admiralty issues a new edition of the *Vocabulary Signalling Book* in three volumes: General Signals, Manoeuvring Signals (with diagrams – see Bigot de Moroques *Tactiques* of 1763) and a General Vocabulary. This remained in force until the beginning of the 20th century.

15

Continuity and Change: Numeral flags in use 1778-1900

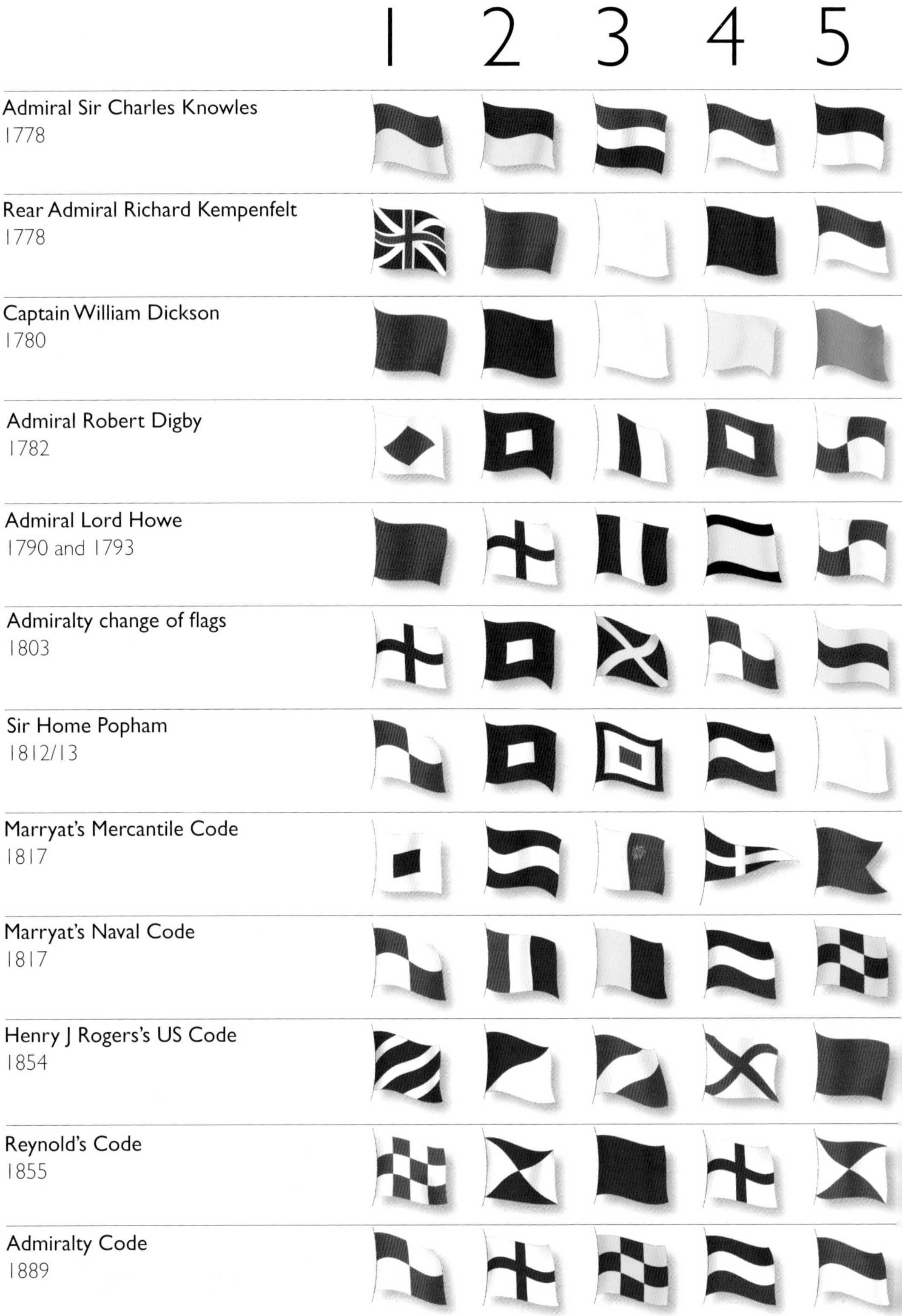

	1	2	3	4	5
Admiral Sir Charles Knowles 1778					
Rear Admiral Richard Kempenfelt 1778					
Captain William Dickson 1780					
Admiral Robert Digby 1782					
Admiral Lord Howe 1790 and 1793					
Admiralty change of flags 1803					
Sir Home Popham 1812/13					
Marryat's Mercantile Code 1817					
Marryat's Naval Code 1817					
Henry J Rogers's US Code 1854					
Reynold's Code 1855					
Admiralty Code 1889					

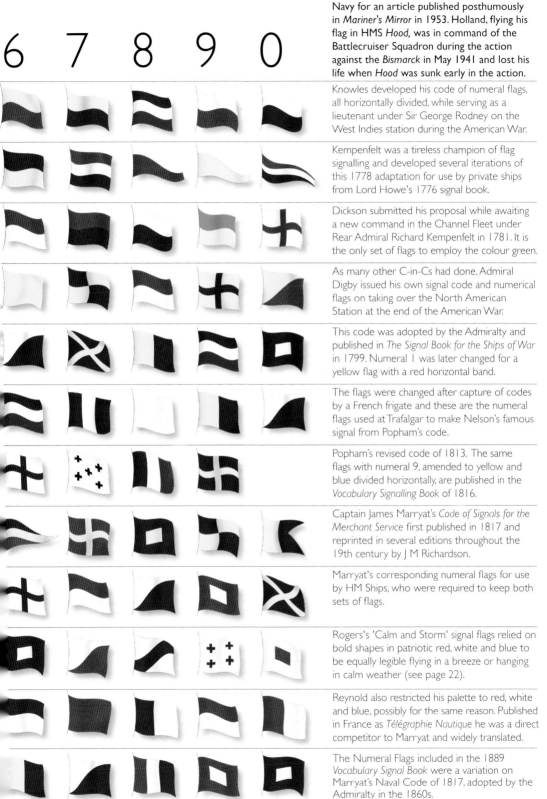

6 7 8 9 0

This chart acknowledges the work of Vice Admiral Lancelot Holland RN who drew a similar chart of flags in use by the Royal Navy for an article published posthumously in *Mariner's Mirror* in 1953. Holland, flying his flag in HMS *Hood*, was in command of the Battlecruiser Squadron during the action against the *Bismarck* in May 1941 and lost his life when *Hood* was sunk early in the action.

Knowles developed his code of numeral flags, all horizontally divided, while serving as a lieutenant under Sir George Rodney on the West Indies station during the American War.

Kempenfelt was a tireless champion of flag signalling and developed several iterations of this 1778 adaptation for use by private ships from Lord Howe's 1776 signal book.

Dickson submitted his proposal while awaiting a new command in the Channel Fleet under Rear Admiral Richard Kempenfelt in 1781. It is the only set of flags to employ the colour green.

As many other C-in-Cs had done, Admiral Digby issued his own signal code and numerical flags on taking over the North American Station at the end of the American War.

This code was adopted by the Admiralty and published in *The Signal Book for the Ships of War* in 1799. Numeral 1 was later changed for a yellow flag with a red horizontal band.

The flags were changed after capture of codes by a French frigate and these are the numeral flags used at Trafalgar to make Nelson's famous signal from Popham's code.

Popham's revised code of 1813. The same flags with numeral 9, amended to yellow and blue divided horizontally, are published in the *Vocabulary Signalling Book* of 1816.

Captain James Marryat's *Code of Signals for the Merchant Service* first published in 1817 and reprinted in several editions throughout the 19th century by J M Richardson.

Marryat's corresponding numeral flags for use by HM Ships, who were required to keep both sets of flags.

Rogers's 'Calm and Storm' signal flags relied on bold shapes in patriotic red, white and blue to be equally legible flying in a breeze or hanging in calm weather (see page 22).

Reynold also restricted his palette to red, white and blue, possibly for the same reason. Published in France as *Télégraphie Nautique* he was a direct competitor to Marryat and widely translated.

The Numeral Flags included in the 1889 *Vocabulary Signal Book* were a variation on Marryat's Naval Code of 1817, adopted by the Admiralty in the 1860s.

There are more than 200 individual flag and pennant illustrations in this book and with the exception of a very small number which use green, they are made up of only four colours: red, yellow, blue, and black plus white. Size alone does not guarantee legibility, not least when there is no wind to reveal the design or, as we shall see later, smoke and mist intervene. Much rests on distinctions of colour and contrast.

The new doctrines of fleet tactics and the introduction of a structured code by which manoeuvres could be ordered from the mid-17th century led to an exponential increase in the numbers of new flags. Many were introduced by individual admirals and squadron commanders and although four of the flags that we recognise today in the International Code have their origins before the turn of the 18th century, more would be discarded with better understanding of colour, contrast and the effects of different conditions at sea. If you look at these pages and half close your eyes, you will quickly discover what works and what doesn't.

Before Lord Howe's 1793 *Signal Book for the Ships of War* brought to an end the competitive scramble for perfection, there was no shortage of opinion: 'chequered flags should be abolished.' (Young – he had a point with the red and blue); 'vertical stripes preferable to horizontal' (Kempenfelt); '...preference for two and three striped flags' (Howe).[1] When Sir Home Popham revised and updated his *Marine Vocabulary* in 1812 he was warmly congratulated by Viscount Hood for the selection of his flags '...with great circumspection.'[2]

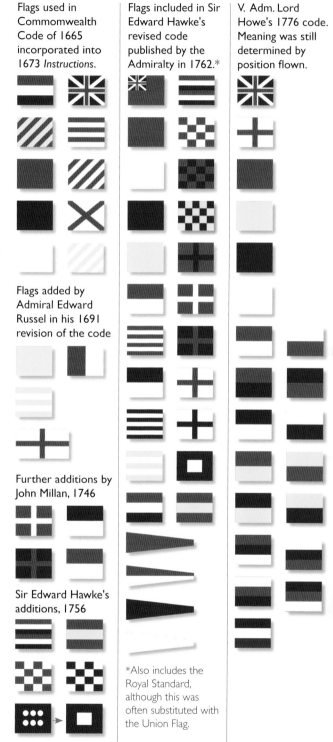

Flags used in Commomwealth Code of 1665 incorporated into 1673 *Instructions*.

Flags added by Admiral Edward Russel in his 1691 revision of the code

Further additions by John Millan, 1746

Sir Edward Hawke's additions, 1756

Flags included in Sir Edward Hawke's revised code published by the Admiralty in 1762.*

V. Adm. Lord Howe's 1776 code. Meaning was still determined by position flown.

*Also includes the Royal Standard, although this was often substituted with the Union Flag.

R. Adm. Richard Kempenfelt's 1778 10^2 matrix of numeral flags for private ships.	Howe/Kempenfelt's numerical flags for 16^2 matrix allowing 256 two-flag hoists, 1778.	Kempenfelt's revised 10^2 matrix* matrix for both flagships and private ships, 1781.	Adm. George Rodney's 1782 additions to Lord Howe/Kempenfelt's 1778 codes.

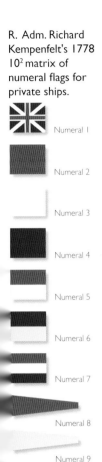

Numeral 1
Numeral 2
Numeral 3
Numeral 4
Numeral 5
Numeral 6
Numeral 7
Numeral 8
Numeral 9
Numeral 10

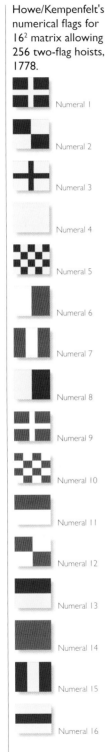

Numeral 1
Numeral 2
Numeral 3
Numeral 4
Numeral 5
Numeral 6
Numeral 7
Numeral 8
Numeral 9
Numeral 10
Numeral 11
Numeral 12
Numeral 13
Numeral 14
Numeral 15
Numeral 16

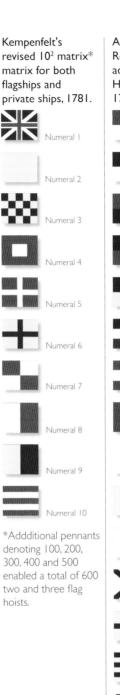

Numeral 1
Numeral 2
Numeral 3
Numeral 4
Numeral 5
Numeral 6
Numeral 7
Numeral 8
Numeral 9
Numeral 10

*Addditional pennants denoting 100, 200, 300, 400 and 500 enabled a total of 600 two and three flag hoists.

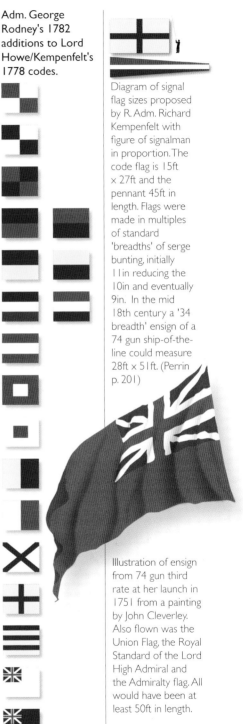

Diagram of signal flag sizes proposed by R. Adm. Richard Kempenfelt with figure of signalman in proportion. The code flag is 15ft × 27ft and the pennant 45ft in length. Flags were made in multiples of standard 'breadths' of serge bunting, initially 11in reducing the 10in and eventually 9in. In the mid 18th century a '34 breadth' ensign of a 74 gun ship-of-the-line could measure 28ft × 51ft. (Perrin p. 201)

Illustration of ensign from 74 gun third rate at her launch in 1751 from a painting by John Cleverley. Also flown was the Union Flag, the Royal Standard of the Lord High Admiral and the Admiralty flag. All would have been at least 50ft in length.

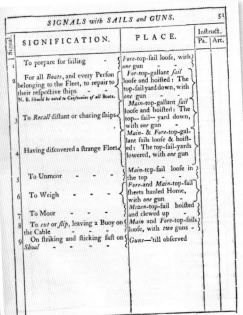

Signal	SIGNIFICATION.	PLACE.	Instruct. Pa.	Art.
1	To prepare for failing	Fore-top-fail loofe, with one gun		
2	For all *Boats*, and every Perfon belonging to the Fleet, to repair to their refpective fhips. N. B. *Should be noted to Coxfwains of all Boats.*	For-top-gallant *fail* loofe and hoifted : The top-fail yard down, with one gun		
3	To *Recall* diftant or chacing fhips	Main-top-gallant *fail* loofe and hoifted : The top-- fail-- yard down, with *one* gun		
4	Having difcovered a ftrange Fleet	Main- & Fore-top-gallant fails loofe & hoifted : The top-fail-yards lowered, with *one* gun		
5	To Unmoor	Main-top-fail loofe in the top		
6	To Weigh	Fore-and Main-top-fail fheets hauled Home, with *one* gun		
7	To Moor	Mizen-top-fail hoifted and clewed up		
8	To *cut* or *flip*, leaving a Buoy on the Cable	Main and Fore-top-fails loofe, with *two* guns		
9	On ftriking and fticking faft on Shoal	*Guns*—'till obferved		

Page from 1799 printing of the *Signal Book for the Ships of War* with instructions for distance signals using a combination of sails and guns, illustrated below.

Fore top-gallant loose and tops'l yard lowered, one gun fired: every person belonging to the fleet to repair to their respective ships.

Main top-gallant loose and tops'l yard lowered, one gun fired: recall distant or chasing ships.

Main and fore top-gallants loose, tops'l yards lowered, one gun fired: to report discovery of a strange fleet.

Fore and main tops'l sheets hauled home, one gun fired: to weigh anchor.

Main and fore tops'ls loose, two guns fired: to cut or slip, leaving a buoy on the cable.

But however well chosen the colours and their arrangement, flags could only go so far. Their visibility was easily disrupted by sails, smoke and poor visibility. Above all, flag signalling was limited by distance.

One of the first known uses of sails as a signalling device over long distances dates from the 14th century with a specific mention of sails in Amadeo VI of Savoy's instructions to his fleet in 1366.[3] We have a reference in Nathaniel Boteler's *Dialogues* in 1627 and a page of instructions in Lord Howe's second *Signal Book for the Ships of War* issued in 1790, shown here. Although these are fairly mundane in nature, distance signalling with sails and shapes enabled fleets concealed below the enemy's horizon to stay in touch with scouting frigates.

Accounts of signals passed on the eve of Trafalgar from Nelson's inshore scouts, led by the frigate *Euryalus*, and the chain of ships in his advance squadron to the fleet nearly fifty miles off Cadiz, give a vivid impression of distance signalling in action.[4] The signal hoist 370 'Enemy ships are coming out of port' was passed from ship to ship but, at extreme range towards the end of the chain between HMS *Mars* (74) and HMS *Bellerophon* (74), was substituted with the equivalent distance signal: a flag, a pennant and a ball at the fore, main and mizzen mastheads. By 0930 on the morning of the 20th October, Nelson had the confirmation he was waiting for, before the first of Villeneuve's fleet had actually left the safety of Cadiz.

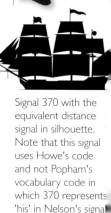

Signal 370 with the equivalent distance signal in silhouette. Note that this signal uses Howe's code and not Popham's vocabulary code in which 370 represents 'his' in Nelson's signal on the morning of 21st October 1805.

23 miles

From the tops'l yard of a 74-gun ship-of-the-line, the sea horizon would be 11.5 miles. Signals could be made over considerable distances.

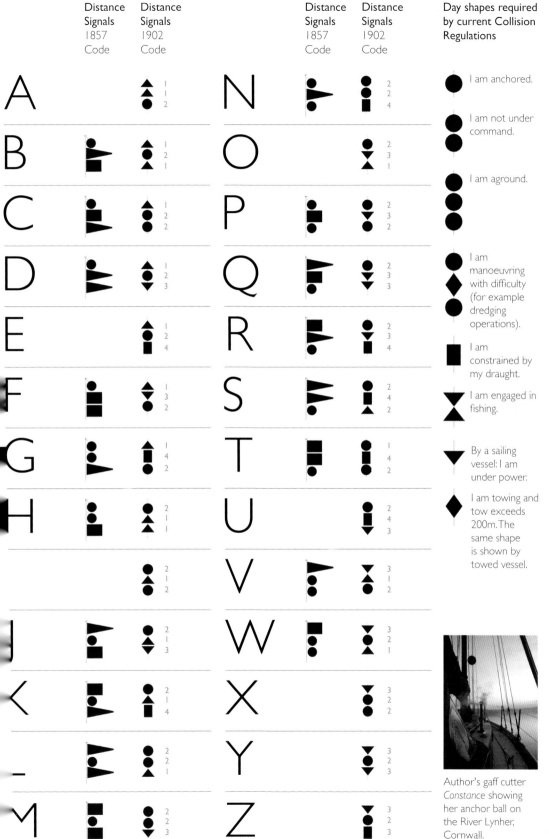

	Distance Signals 1857 Code	Distance Signals 1902 Code			Distance Signals 1857 Code	Distance Signals 1902 Code		Day shapes required by current Collision Regulations
A		1 1 2		N		2 2 4		● I am anchored.
B		1 2 1		O		2 3 1		● ● I am not under command.
C		1 2 2		P		2 3 2		● ● ● I am aground.
D		1 2 3		Q		2 3 3		● ◆ ● I am manoeuvring with difficulty (for example dredging operations).
E		1 2 4		R		2 3 4		■ I am constrained by my draught.
F		1 3 2		S		2 4 2		⧖ I am engaged in fishing.
G		1 4 2		T		1 4 2		▼ By a sailing vessel: I am under power.
H		2 1 1		U		2 4 3		◆ I am towing and tow exceeds 200m. The same shape is shown by towed vessel.
I		2 1 2		V		3 1 2		
J		2 1 3		W		3 2 1		
K		2 1 4		X		3 2 2		
L		2 2 1		Y		3 2 3		
M		2 2 3		Z		3 2 3		

Author's gaff cutter *Constance* showing her anchor ball on the River Lynher, Cornwall.

Calm before the Storm: When the wind doesn't blow

Moments at sea when there is insufficient wind to extend a flag are rare but in the days of sail when lack of wind also denied forward movement to the ship, flag signalling presented different challenges.

Alongside the scramble to develop and market code flag systems throughout the 19th century there were several proposals for calm-weather alternatives. One such came from Captain Francis Liardet RN who was forced to retire from active service after being partially blinded in a cannon explosion in 1841 and turned his attention to the problem. His proposal, published in 1849, uses ten numeral balls and four additional colour-ways to make coded signals in groups of two and three.

Competing with Liardet was Baltimore telegraph engineer Henry J Rogers whose 'American Calm and Storm Signals' were designed to work both extended in a breeze and hanging vertically on the halyard. In the Preface to his 1855 signal book, he highlights the shortcomings of other systems with the consequence that '...coloured balls, cones, cylinders and similar geometrical figures are resorted to... and when two or more balls or cones have to be elevated... the halyards become twisted and cause delay and confusion.'[1]

It was a competitive business to furnish growing merchant fleets with reliable and safe signalling. Another, the Marryat/ Richardson shape system, is shown right.

Above and far right: Captain Francis Liardet's proposed 'Signals to be used in Calm Weather' published in his *Professional Recollections on Points of Seamanship* in 1849. An original watercolour draft of his calm weather signals is held by the National Maritime Museum (EQA 0198.3).

Right: Numeral flags 2 and 3 from Henry J. Rogers's 'Calm and Storm Signals' (see pages 16 and 17) demonstrate distinctive shapes when hanging vertically.

22

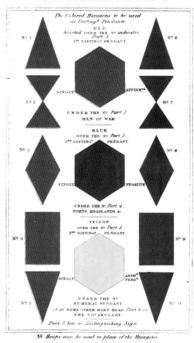

Liardet's calm weather signal 'Weigh anchor'.

Geometric numerical signals from Marryat/ Richardson's 1854 code.

Ensigns and Etiquette: Knowing who's who

The opening chapters of W G Perrin's 1922 study of British Flags first trace the origin of the flag as a national device. In the case of England, he goes on to examine the competing attibutions to several saints, among them St Cuthbert, St Edmund and Edward the Confessor, for the provision the central motif of what would much later become the Union flag. He concludes that by the time King Edward I rallied his armies to resist Welsh claims for independence in 1277 it was most likely the cross of St George under which they marched.

When, some seventy years later in 1346, his grandson King Edward III claimed the divine intercession of St George in his victory over the French at the Battle of Crécy, there was no question of '...the dethronement of Edward the Confessor from the position of "patron saint" of England and the definitive substitution of St George in his place.'[1] The dedication to St George of a new collegiate chapel at Windsor dates from the same moment in our history.

For Scotland, the attribution of a national motif to a particular saint was more straightforward, with St Andrew the only contender. Perrin points out, however, that there is no direct evidence of the cross-saltire being adopted as a

national ensign before the 14th century and then with some variety in the colour against which the white cross is displayed with black, red and yellow (the Stuart colours) competing with the more usual blue.[2]

The accession of King James VI of Scotland to the English throne raised the question of how the two established crosses could be equitably combined to project the authority of the new kingdom, particularly for ships at sea.

A royal proclamation of April 1606 calls upon '... our Subjects of South and North Britain, Travelling by Sea... shall bear in their maintop the Red Cross, commonly called St George's Cross and the White Cross, commonly called St Andrew's Cross joined together according to a form made by our heralds...'[3] In addition, subjects of South Britain were to wear the Red Cross and those of North Britain the White Cross in their foretop. But whatever exact form the heraldic union took (for evidence of the original design was destroyed by fire in 1618), it did not please the shipmasters of Scotland who objected to the division of the saltire by the cross of St George and resisted its use until the formal Acts of Union between the two countries in 1707.

The most likely union of the two crosses, though opinions vary as to the intended width of the fimbriation, or white border, to the St George' cross which some evidence suggests was much wider.

The quartered version of the Union Flag is sometimes referred to as the Commonwealth Flag. It is said to have been used by the fleet sent to return Prince Charles and the Duke of Buckingham from Spain in 1623 but a contemporary painting of the event shows no evidence of it.[4]

Royal Standard, early 16th century, flown by the Lord High Admiral of England.

Ensigns of the Elizabethan Navy. Both were flown during Lord Howard's 1596 expedition to Cadiz.

Ensigns of the Jacobean Navy.

Despite Scottish reluctance to fly it, at the death of King James I in 1625, it is found in a list of flags and banners for display at the King's funeral, referred to for the first time as the Union Flag.

Apart from the different flags flown at the foretop, there was nothing to distiguish merchant ships from those of the new king's Navy Royal. A further proclamation of 1634 required English and Scotish merchantmen to desist from flying the Union Flag – possibly no hardship for the Scots, but perhaps an indication that the Navy had begun to feel undermined at home while threatened by the combined fleets of France and the Dutch Republic.

The Civil War and the execution of King Charles I in 1649 changed everything; the dynastic link with Scotland was broken and the Navy once again reverted to the flag of St George. When the passing of the Navigation Act of 1651 threatened Dutch hegemony over world trade, war threatened and the Commonwealth's Generals-at-Sea determined on a new flag for the commander-in-chief to replace the Royal Standard. In place of the Scotish saltire the new standard brought together the cross of St George and the harp of Cromwell's newly-conquered Ireland. There are variants of this flag but evidence of at least two paintings from the First Dutch War shows it used as both a bowsprit jack (see below left) and as a masthead standard.[5]

With Scotland politically re-united with England in 1654, the Union Flag was reinstated but still denied to merchantmen. Yet another proclamation of 1674 forbade the use of 'any Jack made in imitation'[6] in attempts to gain the same privileges in foreign ports as those enjoyed by ships of the newly named Royal Navy. But by the Acts of Union of 1707 the Red Ensign with the Union Flag in the upper canton was formally adopted by both the Royal Navy and Mercantile Marine.

Union with Ireland came into effect on 1st January 1801, from which date the red saltire of St Patrick was joined with that of St Andrew. Similar debates as in 1606 over the relative widths of the two saltires and their borders were eventually settled, giving us our national flag and, from 1864, the three separate ensigns in use to this day.

Detail of painting by Jan Beerstraaten showing General-at-Sea George Monck's *Resolution* flying the St George cross with the Irish harp from the bowsprit jackstaff at the Battle of Scheveningen on 10th August 1653 during the First Dutch War. A larger section of the painting is shown in the background to the previous page.

Rijksmuseum, Amsterdam [CC0 1.0]

17th century mercantile ensigns of England and Scotland.

1653 standard combining the cross of St George with the Irish harp.

Ensigns of the Blue and White Squadrons. In 1702, the St George's cross was added to the white ensign to avoid confusion with the plain white French ensign.

Devotees of Patrick O'Brian's Aubrey novels will be familiar with Flag Officers of the Red, the White and the Blue; but which has precedence, from which masts were which flags to be flown and when did the order of the colours change?

As fleets grew in size and tactics moved from *mêlée* battles at close quarters to fleet engagements in line-of-battle, new forms of organisation were needed with fleets divided into smaller, more manageable squadrons. One of the earliest records of a squadron formation is that proposed by King Henry VIII's Lord Admiral John Dudley in 1545, in which he specifies a distinct hierarchy of flags ascribing an order of precedence to the masts from which they were to be flown.[7] The distinctions between the squadrons can be tabulated as follows:

	Foretop	Maintop	Mizzen
Lord Admiral			
Admiral of the Van			
Admiral of the Wing			

The private ships of each squadron flew a single St George's flag from main, fore and mizzen tops respectively. By the time Sir Edward Cecil gathered his fleet for the disastrous expedition to Cadiz in 1625, there was a clear instruction on the division of the fleet into three squadrons, each under three admirals with, for the first time, the colours red, blue and white assigned to the squadrons in descending order of seniority. Sources differ on the positions of flags but an expedition two years later under the Duke of Buckingham intended to capture the Isle de Ré gives a clearer indication of how the squadrons might have been identified:

	Foretop	Maintop	Mizzen
Lord High Admiral			*
V. Adm			
R. Adm			
Adm. of the Blue			
V. Adm			
R.Adm			
Adm. of the White			
V Adm			
R. Adm			

* senior Admirals would fly the Union flag

Private ships within the squadrons wore a pennant matching their Flag Officers and at the same masthead, though at this date there is no record of corresponding ensigns being worn. With a change of colour precedence to red, white and blue in March 1653, this squadronal structure remained in place, with the adoption of corresponding ensigns, until colour squadrons were abandoned in 1864.

Promotion path for Flag Officers to 1805.

- Rear Admiral of the Blue
- Rear Admiral of the White
- Rear Admiral of the Red
- Vice Admiral of the Blue
- Vice Admiral of the White
- Vice Admiral of the Red
- Admiral of the Blue
- Admiral of the White
- Admiral of the Fleet

The ensigns shown above are those in effect from 1702 to 1801. After the Battle of Trafalgar in 1805 an additional rank of Admiral of the Red was created, the highest rank attainable until 1862 when more than one Admiral of the Fleet was permitted. Two years later the squadronal system was abandoned as having no further useful tactical purpose.

Ensign of the East India Company, early 19th century.

Red Ensign reserved for British merchant ships from 1864. Note the change to 2:1 ratio.

Blue Ensign of the Royal Naval Reserve and Fleet Auxiliaries denoted by the gold anchor.

White Ensign flown only by HM ships and, from 1859, by members of the Royal Yacht Squadron.

Nelson Confides: Decoding the Trafalgar Signal

Much has been written about Admiral Lord Nelson's famous signal as his two columns made their slow downwind approach towards the combined French and Spanish fleets off Cape Trafalgar on 21st October 1805. What doesn't get discussed so often is the Admiral's intended message. We can be fairly certain, not least from Lieutenant Pasco's own recollections of his instructions, that Nelson had asked for the word 'confides' which was, with his agreement, substituted with 'expects'; the former requiring to be spelled out while the latter was coded as a three-flag hoist in Popham's *Vocabulary Signalling Book*. But was his intention to signal 'England confides...' or, as historian David Howarth suggests in his 1969 *Trafalgar, The Nelson Touch*[1] the less jingoistic intimation 'Nelson confides...'?

Gillray cartoon lampooning Lady Hamilton as *Dido in Despair* when Nelson sailed for Copenhagen.

Despite the lampooning of Nelson and his relationship with Emma Hamilton in the press and popular prints, there is no doubting the mutual affection, respect and reciprocated confidence he inspired in the men under his command. There is only anecdotal evidence that he proposed the opening words 'Nelson confides' in a conversation with Captain Blackwood of the frigate *Euryalus*, who was still aboard *Victory*

Telegraph or Cipher Flag

2

5

3

England

2

6

9

expects

8

6

3

that

2

6

1

every

at the time. Whether he changed his mind or was persuaded otherwise we don't know. But from the testimony of Nelson's own words, writing to Admiral Lord Howe acknowledging both Howe's congratulations on his success at the Battle of the Nile in August 1798 and the effective deployment of Howe's new signal flags, he says 'I had the happiness to command a Band of Brothers... each knew his duty'. [2] Surely it would be a similar sentiment of quiet confidence that he personally wished to express directly to his captains and men before the opening shots were fired at Trafalgar? The signal immortalised in heroic memory doesn't quite do that.

How was the signal received? From a mix of anecdotal evidence and records from logbooks of ships in the fleet, it seems likely, despite the efforts of the repeating frigates deployed between the two columns, that it was seen by only a handful of ships. Responses varied from muttered oaths about duty to Vice Admiral Collingwood's variously reported remark 'I wish Nelson would stop signalling. We know well enough what to do.'[3] Nevertheless, when the full signal was read to him, Collingwood, in the van of the leeward division, approved it warmly as did the 'Tars of the Tyne' as he liked to address his men aboard the *Royal Sovereign*. According to an account published in the *Navy and Army Illustrated* of 1896, they greeted the message with a 'burst of cheering'.[4]

But whatever its immediate effect and irrespective of Nelson's original intention, the death of the Admiral at the moment of victory soon endowed it with mythic status; often misquoted

Royal Museums Greenwich (PW 3874)

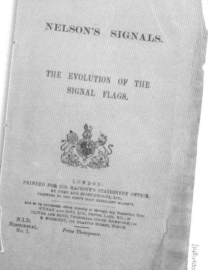

4	
7	
1	
man	
9	
5	
8	
will	
2	
Repeat (2)	
0	
do	
3	
7	
0	
his	

and frequently re-appropriated to this day. While documents of record give us some certainty over the words encoded: "England expects that every man will do his duty", shown in the twelve hoists on these pages and above in an Australian postcard from WWI, versions omitting and adding words abound, with the opening words 'England Expects' frequently used as shorthand for the full signal. And, if there is confusion over the exact wording, there has, at least until the painstaking work undertaken by Lt Colonel W G Perrin in 1908, been equal confusion over the flags used to make the signal.

Perrin, upon whose scholarly 1922 study *British Flags: Their Early History and Their Development at Sea* much of the timeline of the previous section is based, was appointed Admiralty Librarian in 1908 during Admiral 'Jackie' Fisher's first tenure as First Sea Lord. It was his research published in *Nelson's Signals: The Evolution of The Signal Flags* (right) in the same year that revealed the errors in earlier depictions of the

4	
d	
2	
1	
u	
1	
9	
t	
2	
4	
y	
End of signal	

The flag hoists on these pages depict Nelson's signal made using Popham's Marine Vocabulary of 1803. Each word is encoded numerically in a three-flag hoist with the exception of 'duty' which had to be spelled out. Each letter of the alphabet was represented by numerals 1 to 25, with 'I' and 'J' combined (9) and 'V' (20) and 'U' (21) reversed.

The signal, most likely hoisted in succession on the mizzen flag halyards to afford greatest visibility to ships astern, was quickly followed by Nelson's signature call for 'Close Action' – signal 16 in Popham's code. It remained aloft until shot away.

souvenir postcard of HMS *Victory* published for the Trafalgar Centenary in 1905. The flag hoists are taken from the superceded 1799 code.

NELSON'S SIGNALS.

THE EVOLUTION OF THE SIGNAL FLAGS.

LONDON
PRINTED FOR HIS MAJESTY'S STATIONERY OFFICE,
BY EYRE AND SPOTTISWOODE, LTD.,
PRINTERS TO THE KING'S MOST EXCELLENT MAJESTY.

N.I.D.
HISTORICAL
No. 1.   Price Threepence.

Trafalgar signal. These had been based on the assumption that the 1799 code was still in use in October 1805. In fact, as Perrin discovered, the code had been changed and manuscript amendments issued to the fleet after the capture of the 1799 code book in April 1803.

There can be few military signals that have embedded themselves so deeply into national popular culture or, in different conflicts, been reprised to express a similar sentiment – even in a single flag.

to emulate with a single flag at Tsushima and flag Zulu still flies to this day from the upper yard of his flagship *Mikasa*, now a museum ship in Yokosuka.

Code flag 'Z' hoisted aboard Admiral Tōgō's flagship before battle was joined at Tsushima. It still flies today on the Mikasa, now a museum ship (below).

Japanese Admiral Tōgō Heihachirō, before engaging the Russian fleet at the Battle of Tsushima in May 1905, ordered the single hoist of International Code flag 'Z'. By pre-arrangement this was to be interpreted as 'The rise or fall of the Empire depends upon this battle; everyone will do his duty with utmost efforts'. Tōgō had trained in Britain, joining the Training Ship HMS *Worcester* on the River Thames at Greenithe in 1872. Gunnery training was carried out aboard HMS *Victory* at Portsmouth where, in his biography of the admiral, Jonathan Clements records the speed with which the young Tōgō worked out the hoists of Nelson's Trafalgar signal flying from *Victory*'s masts.[5] The signal made a lasting impression that he was

Had Popham's code enabled Nelson's originally intended intimacy of confidence rather then the mandatory 'expectation' signalled to the fleet, would the signal have had such a lasting impact?

And if, as David Howarth has argued,[6] the signal had been made with the personal appeal of Nelson rather than the substituted 'England', would it so readily have been pressed into service for everything from recruitment posters in both World Wars (above and right) to a history of the England football team? Or perhaps it would have been re-invented to meet the need.

Commercial Codes

The publication in 1857 of the first Commercial Code overlapped with later editions of the Marryat Code, with many mariners preferring the latter. HM ships on the East Indies station continued to use Marryat's code to communicate with merchant ships and there is anecdotal evidence of the code still in use in the 1890s,[1] by which time the original *Commercial Code* published by the British Board of Trade in 1857 was under review by the Washington International Maritime Conference of 1889. This resulted in a revised code now with a full alphabet published in 1897 and officially adopted in January 1902.

The code was again revised in 1931 and published as the *International Code of Signals* in seven languages with effect from January 1934. Rectangular flags were used for all 26 letters of the alphabet and numeral pennants introduced together with three triangular substitute flags and an answering pennant/code flag, shown here on the cover of the author's 1961 edition of *Brown's Signalling*[2] that no cadet could be without. This is the code that is still in use today.

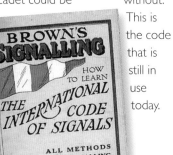

	18 flag code of 1857	1901 Code (effective 1st January 1902)	1931 Code (effective 1st January 1934)	Individual flag meanings (1934)
A				I am undergoing speed trials.
B				I am taking in or discharging explosives.
C				Affirmative.
D				Keep clear of me – I am manoeuvering with difficulty.
E				I am directing my course to starboard.
F			*	I am disabled. Communicate with me.
G				I require a pilot.
H				I have a pilot on board.
I				I am directing my course to port.
J				I am going to send a message by semaphore.
K			*	You should stop your vessel instantly.
L			*	Stop, I have something important to comminucate
M				I have a doctor on board.

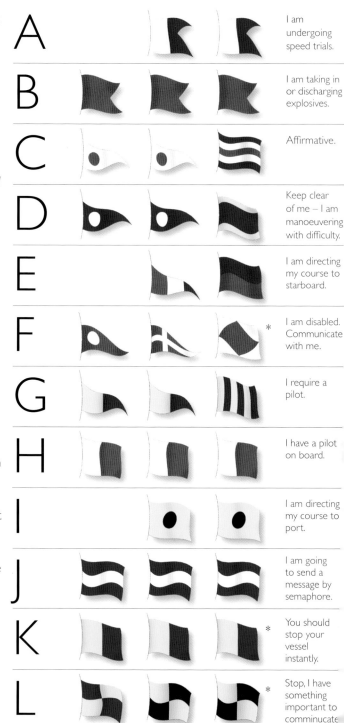

18 flag code of 1857	1901 Code (effective 1st January 1902)	1931 Code (effective 1st January 1934)	Individual flag meanings (1934)	Numerals, pennants and substitute flags introduced in 1934 code.	
N				Negative.	1	Note that flags P, T and X followed by a numeral hoist indicate position, time or a true bearing respectively. As an example, the hoist shown below would be read as latitude 49° 50' N, although the N would often be omitted, it usually being obvious which hemisphere is referred to.
O			*	Man overboard.	2	
P			*	Vessel about to sail or your lights are burning badly.	3	
Q				My vessel is healthy and I request free pratique.	4	
R			*	The way is off my ship and you may feel your way past.	5	
S				My engines are going full speed astern.	6	
T				Do not pass ahead of me.	7	
U			*	You are standing into danger.	8	
V			*	I require assistance.	9	
W			*	I require medical assistance.	0	
X				Stop carrying out intentions and watch for my signals.	First Substitute	
Y				I am carrying mails.	Second Substitute	
Z			*	To be used to address or call shore stations.	Third Substitute	* Single letter codes marked with an asterisk could also be signalled by morse lamp.

31

Norfolk Museums Service (Time and Tide Museum)

Though the International Code has been adotped throughout the world, we should not underestimate the impact that some of the earlier proposals had when first introduced. While the Royal Navy had settled on Popham's *Vocabulary Signalling Book* in 1816, there was no universal equivalent for merchant shipping until the publication of Frederick Marryat's *Code of Signals for the Merchant Service*. Published in 1817, it was the first of many competing proposals (see Timeline pages14 and 15) and one of two that deserve attention here.

The other is the *Telegraphic Dictionary and Seaman's Signal Book* first developed by Henry J Rogers in Baltimore in 1845. This became his *American Code of Signals* which was adopted by both the Merchant Marine and the US Navy with an endorsement from the Secretary of the Navy J C Dobbin in April1855 urging Commanding Officers to '...embrace every opportunity to familiarise the Service with the use of these signals'.[3]

Both codes used numeral flags only with Marryat's employing six additional flags to qualify the meaning, two examples of which are shown here: a ship's identifying signal letters and a four-flag hoist preceded by the 'Rendezvous' flag to indicate a port or destination.

4
0
7
2

Marryat's First Distinguishing Pennant followed by numerals 4072 of the British barque Frederica of Yarmouth (above) painted in 1865.

7
0
8
6

Four numerals preceded by the 'Rendezvous' flag indicates a geographical location, in this case Braye Harbour on Alderney.

A letter from a ship master to the *Nautical Magazine* in April 1840 testifies to the value of Marryat's system, describing how he was warned of reefs and a safe course to steer in the Torres Strait by another ship. The letter goes on '...by means of Marryat's *Code of Signals* any telegraphic communicatioin whatever may with facility be transmitted'.[4] He ends by observing that if shipowners '...did properly appreciate the value of these signals, no ship would be without them'. It is fortunate that both ships in the exchange were equipped with a set of flags and Marryat's code. Clearly not all were.

The six sections and 391 pages (Richardson edition, 1861) of Marryat's code book list permutatioms of four-flag hoists; Rogers's *American Code* used combinations of two, three, four and five numeral flag hoists and listed more than 60,000 closely printed individual signals. Subjects ranged from the purely practical: time signals, pilotage and navigational hazards to intelligence on the price of stocks. Signal 247 shown left, for example, represents a one-sixteenth rise in cotton prices.[5] Shown below is a more mundane exchange which echoes the title of this book.

6
3

3
6

1

Two hoists in Rogers's code: 'Where are you bound?' with the answer 'New York'.

9

5

32

Alfred Jensen, 1893

The *Commercial Code* published by the Board of Trade in 1857 broke with the all-numerical codes, using 18 alphabetic flags in permutations of two-, three- and four-flag hoists. With no vowels, there was little scope for spelling out words, but it was the beginning of a universal system that gradually replaced the Marryat code. A note on the title page of the 1873 supplement of British ship codes (some 18,700 vessels) expressly urges that 'This Code is used on board HM ships and that Marryat's code is no longer supplied to HM ships'.[6] A later advertisement offers to provide the necessary flags to fully comply with the International Code for twelve shillings and warned that insurers would require t '...as a means of seaworthiness'.

R

J

P

T

The German barque *Pisagua*, painted in 1893, flies her signal letters in the International Code. Signal letters for German ships later changed to start with 'D', the initial 'R' reserved for Russia and, later, the Soviet Union.

Answering Pennant

The following message is spelled out alphabetically.

E

E

Y

Two-flag hoists were reserved for urgent signals relating to navigation and safety at sea, and remain so to this day. The hoist shown here requests the depth of water on the bar.

G

J

L

Three-flag hoists AAA to ADV were reserved for compass bearings and time; ADW-AHB for a model verb and qualifying phrases; AHX onwards were allocated to the General Code. This hoist warns 'I have had fog whilst passing through ice area'.

A

1st Substitute

Three-flag hoist indicating a relative bearing. AAR is 30° to port.

R

IWM SP 1310 [PD-UKGov]

G

Q

P

R

G

Q

P

R

Four-flag hoists beginning with G were initially reserved for HM ships. Here the light cruiser HMS *Caroline*, now a museum ship in Belfast, is represented in the 1902 International Code, and the 1913 Naval Code.

A

M

T

3rd Substitute

Four-flag hoists beginning with 'A' denote a geographical location. AMTT is Port Said.

British Naval Codes

If the British Admiralty led the way in the adoption of Popham's *Vocabulary Signalling Book* in 1816, by the time the International Code of Signals had been endorsed by the Washington Conference of 1889 as the universal maritime code, the Admiralty was far from settled on a lasting equivalent. Underlying a near continuous process of enquiry and committee deliberation were two opposing schools of thought on what the signal book was for.

The development of flag signalling from the earliest days was inextricably linked to the communication and control of fleet manoeuvres under sail. The transition to steam, with course and speed no longer governed by wind direction, allowed ever more precise instructions, or steam tactics as they became known, to be codified in flag hoists – characterised by Andrew Gordon as 'the *Signal Book* goosestep'.[1] It was this polarisation of centralised control versus individual initiative, that lay at the heart of fierce tactical debates in the Royal Navy throughout the ninetheenth century.

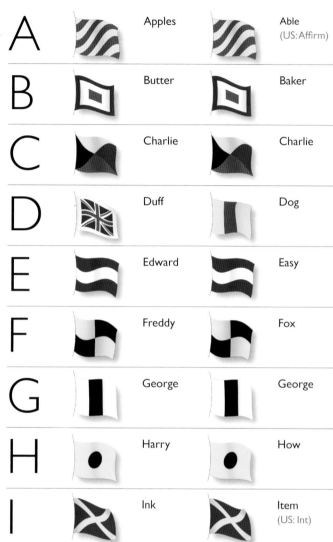

	1915 Alphabetic Flags	1915 Phonetic Alphabet	1937 Alphabetic Flags	1937 Phonetic Alphabet
A		Apples		Able (US: Affirm)
B		Butter		Baker
C		Charlie		Charlie
D		Duff		Dog
E		Edward		Easy
F		Freddy		Fox
G		George		George
H		Harry		How
I		Ink		Item (US: Int)
J		Johnnie		Jig
K		King		King
L		London		Love
M		Monkey		Mike

1st Substitute	
2nd Substitute	
3rd Substitute	
4th Substitute	

	1915 Alphabetic Flags	1915 Phonetic Alphabet	1937 Alphabetic Flags	1937 Phonetic Alphabet		Numeral Flags 1913 and 1937	Numeral Pennants 1913 and 1937*
N		Nuts		Nan (US: Negat)	1		
O		Orange		Oboe (US: Option)	2		
P		Pudding		Peter (US: Prep)	3		
Q		Queenie		Queen	4		
R		Robert		Roger	5		
S		Sugar		Sugar	6		++
T		Tommy		Tare	7		
U		Uncle		Uncle	8		
V		Vinegar		Victor	9		
W		William		William	0		
X		Xerxes		Xray		*No 9 Pennant 1937 Code	
Y		Yellow		Yoke		First Substitute Pennant	
Z		Zebra		Zebra		Second Substitute Pennant	

35

	1913 Signal Book	1937 Signal Book
	Affirmative	Affirmative
	Blue Affirmative	Blue Affirmative
	Negative	Negative
	Preparative	Preparative
	Red Burgee	Battleship
	Blue Burgee	Stationing
	Interrogative	Interrogative
	Answering	Answering
	Guard Pennant	Disposition
	Numerical	Numerical
	Church Pennant	Church Pennant
	Fishery Duty Pennant	Fishery Duty Pennant
	Compass/ Red Pennant	Red Pennant

	1913 Signal Book	1937 Signal Book
	Blue Pennant	Blue Pennant
	White Pennant	White Pennant
	Equal Speed Pennant	Order
	Oblique Pennant	Formation
	Union Flag	Union Flag

Anecdotes abound among signalmen of cryptic exchanges. The two pennant hoist (Church and Interrogative) shown here is reported to have been made by an inexperienced escort captain during WW2 after becoming detached from his convoy and uncertain of his position. Asked what his signal meant when eventually closing with a British warship, he replied 'Oh God, where am I?'[2]

Below: Re-interpretation from Eric Tuffnell's Midshipman's Journal of Nelson's Trafalgar signal using the alphabetic Naval Code of 1905.[3]

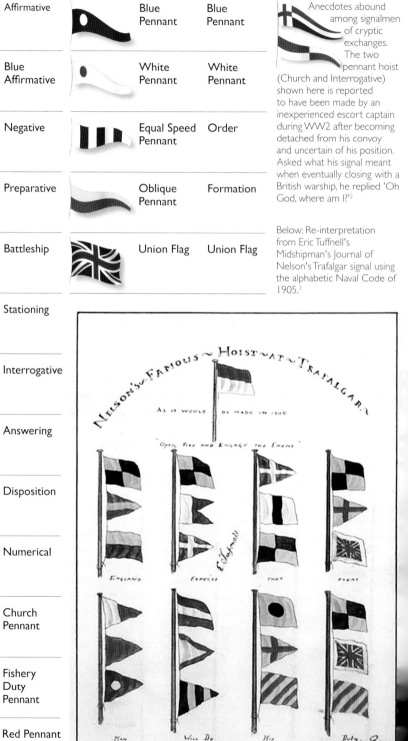

NELSON'S ~ FAMOUS ~ HOIST ~ AT ~ TRAFALGAR.

AS IT WOULD BE MADE IN 1905

"Open Fire and Engage the Enemy"

ENGLAND EXPECTS THAT EVERY

MAN WILL DO HIS DUTY.

1937
Signal Book
Additions

1937
Signal Book
Additions

US Special Flags

Issued to HM Ships for combined fleet signalling with
US Navy General Signal Book, 1944[4]

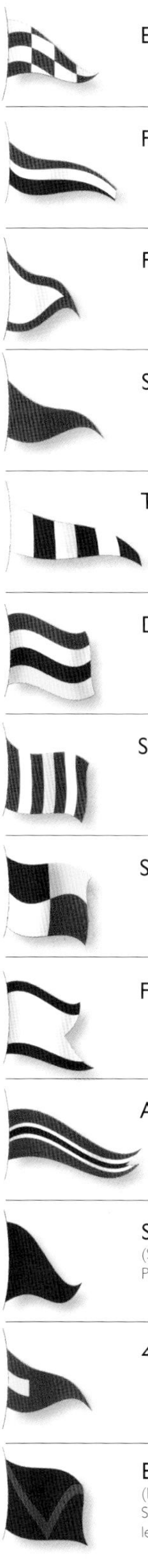

	Aeroplane		Screen	1	EMERG
	Ahead		Squadron/ Flotilla	2	FORM
	Aircraft Carrier		Starboard	3	POSIT
	Astern		Sub Division	4	SPEED
	Battle Cruiser		Submarine	5	TURN
	Black Flag		Bearing	6	DIV
	Blue Flag		Course	7	SECT
	Cruiser		Deployment	8	SQUAD
	Destroyer		Large Black Pennant	9	FLOT
	Division			0	ANS
	Optional				CORPEN (Course Pennant)	SOPUS (Senior Officer Present)
	Port				DEPLOY	4th Repeat
	Red Flag				DESIG	BUS (British/United States, see note left)*

* Used to denote signal
exchanges between British
and US Warships using the
USN General Code Book.

37

'Land's End for Orders': Signalling ship to shore

Before radio enabled shipowners and their agents to communicate directly with ships at sea, the only means of passing instructions to masters, or receiving information on their safe passage, was by visual signal to a coastal signal station. For the last three decades of the 19th century most of the growing number of Lloyd's Signal Stations were connected to the electric telegraph and by Act of Parliament of 1888 Lloyd's were mandated to connect all their stations to the Post Office telegraph network. Within three years, there were some 40 such stations round the coast of the British Isles and 118 established worldwide; by 1923, that number had risen to 134.[1]

Lloyd's primary interest was in the safety of the ships whose insurance risk they bore; for shipowners an equal benefit was the ability to control more precisely the routing and scheduling of their ships with increasing competition for lucrative cargoes. Shipping companies began to offer a 'liner' service with regular departures and scheduled arrivals, though for some cargoes, particularly oil in bulk, discharge ports were often uncertain, dependent on refinery capacity and, not least, the price the oil could fetch. Homeward bound masters would be instructed to proceed to 'Land's End for Orders', at least a week's passage after leaving the Suez Canal, before learning their final destination.

In practice, although there was a Coastguard Station on Gwenap Head, just south of Land's End itself, now part of the National Coastwatch network, ships, whether expecting orders or not, would routinely exhchange signals using the International Code – the only one recognised – with the Lloyd's Signal Stations either on St Mary's or the Lizard. Reports of their passage, whether inbound or outbound would then be passed by telegraph to their owners and published in the daily *Lloyd's List and Shipping Gazette*.

The 1873 Supplement to the International Code of Signals lists 17 coastal stations which did not at that date include the Lizard, although a signal station had been established there in the previous year; in fact two rival groups of Falmouth ship-owners competed for the business eventually taken over by Lloyd's in 1883.[2] There was then a signal station listed at Penzance with the next one to the eastward at Prawle Point. Here the signal station boasted the ability to report directly to the *Shipping Gazette* by 'private telegraph'.[3]

Below: Hand coloured postcard of Lloyd's Signal Station at the Lizard c. 1905. The hoist is probably WCGN, illustrated above, indicating an exchange with an American ship.

Right: The Lizard signal station today, a Grade II listed building and now a private house.

The map on the right shows the distribution of the Signal Stations listed in International Code Books of 1873 and 1923 by which time wireless telegraphy had begun to replace flag signalling for ship-to-shore communication. As with the competition to build a signal station on the Lizard, many stations were built and operated privately though by the turn of the century all were operated by or on behalf of Lloyd's. Some, such as that on Lundy Island, acted both as Lloyd's Signal Stations and Naval Signal Stations and were manned by Royal Navy personnel.

Pre-dating the Lloyd's Signal Stations by nearly a century, the map also shows the extensive network of signal stations established during the Napoleonic Wars, initially to counter the threat from commerce raiders in the English Channel and later the threat of invasion. The first chain of 25 stations, each manned by a naval lieutenant on half-pay and two ratings, was complete from Land's End to Poole Harbour in 1794; within four years a further 47 stations extended the network along the south coast, through the Downs and across the Thames Esturary to Great Yarmouth.[4] A failed French attempt to invade Ireland in 1798 prompted an equally extensive network from Donegal to Dublin Bay, many of them substantial stone structures which survive today. All were in line of sight from each other and their purpose was to watch shipping and relay, using a code of flags, pennants and canvas balls, enemy sightings to the nearest naval C-in-C or directly to frigates at sea.

During the second phase of building, many of the stations were connected to the Admiralty in London by semaphore

••• Napoleonic Signal Stations
■ Coastal Signal Stations 1873
■ Lloyd's Signal Stations 1923

telegraph. One of the first, in January 1796, was at Deal on the Kent coast, also the site of the first of the Lloyd's network which began operating in 1852. But it was not the first signal station. If Deal's function was restricted to passing on shipping intelligence and, from 1855 relaying the 1 pm time signal in synchronisation with the time ball at Greenwich, one of its predecessors, some 300 miles to the northwest,

The Signal Station and Greenwich Timeball on the seafront at Deal.

Signal Tower on Lettermullan Island, Co Galway.

Photo: Stuart Rathbone.
The Construction, Survival, and Use of the Signal Defensible Guard Houses in Connacht and Ulster, PhD Thesis submitted to IT Sligo, February 2020.

View to the north west in painting 'Bidston Lighthouse and Telegraph in 1825' by Robert Salmon.

served both as a signal station and a proud display of Liverpool's trading wealth. Occupying the ridge of Bidston Hill just over three miles to the west of the city, a lighthouse and signal station were established in May 1763 'by order of Liverpool Common Council... at the expense of the Dock duties'.[5]

At an elevation of 230ft (70m), one of the highest points on the Wirral, the signal station could monitor ships approaching Liverpool by both the Rock and Formby channels. Ship owners were warned of the imminent arrival of one of their ships by the flying of a dedicated flag from one of dozens of numbered flagpoles. Each was allocated to either an owner or type of vessel: brigs, snows, men of war and, later, the regular steam packet services between Liverpool and Ireland. But above all they were a reassuring public display of the commerce at the heart of the city's growth.

Writing in 1948, the Liverpool maritime historian Arthur C. Wardle[6] points to adoption of many of these flags as the permanent house colours and liveries of Liverpool shipping companies once the arrival of the semaphore linking Holyhead with Liverpool in 1827 made the signal flags redundant. You don't have to look far down the lists of the Bidston flagpoles, many published with the tide tables as popular guides, to find familiar names: Brocklebank, Holt and Bibby whose red flag (defaced in 1926 with the family crest to avoid confusion with the red flag of Bolshevik Russia) is still in use today. Also high on the list of owners are inescapable reminders of Liverpool's lucrative participation in the triangular trade that filled the city's warehouses with tobacco and sugar. Describing the commerce of the city, William Enfield's 1774 *Essay towards a History of Liverpool* records the names of 105 ships engaged in the African trade for the year 1771, accounting for the shipment of 28,200 slaves.[7]

But the spectacle of Bidston attracted visitors and captured the contemporary imagination, with the Manchester Theatre advertising in 1785 a performance that included '...an exact replica of vessels coming round the Rock; Bidstone Lighthouse; the *Ceremony of the Signals...*'.[8] Not many signal stations enjoyed such popular acclaim.

Images of the signal station and accompanying lists of owners' flagpoles were a popular motif, here depicted on a Liverpool creamware jug of 1793 (above) and in a detail from an English sampler of 1795 now in the Cooper Hewitt Collection of the Smithsonian Museum (right).

40

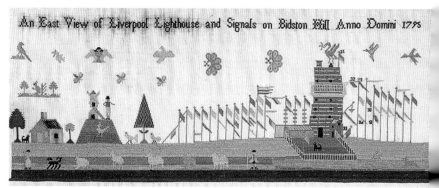

Unintended Consequences: Sending mixed messages

In order to understand how a well-intentioned signal hoist could sow confusion and lead to unintended consequences and even disaster, it is important to follow the not-quite-parallel development of fleet tactics and the signals by which they were communicated.

For more than a century from the Dutch Wars until the loss of the American colonies, naval tactics were governed by the *Articles of Sailing and Fighting Instructions* drawn up under the Commonwealth Code of Generals-at-Sea Monck and Blake, expanded by the Duke of York at the Restoration and polished by Admirals Russell and Rooke at the turn of the 18th century. At their heart was the doctrine of the line-of-battle which, despite the advantage it gave in maximising the firepower of the fleet, many regarded as a tactical straightjacket that impeded initiative and led to inconclusive outcomes. It would be easy to characterise the period as one of sterility and glacial change but Richard Harding in his *Seapower and Naval Warfare 1650-1830* cautions against that temptation, pointing to considerable gains in the operational flexibility and control of large fleets.[1] Sir Julian Corbett in his 1895 analytical study for the Navy Records Society *Fighting Instructions 1530-1816* also concludes that even by mid-century at the hands of Admirals Vernon, Anson and Hawke '...a ripe and sound system of tactics had been reached'.[2]

On pages 18 and 19 we saw the exponential growth in signal flags in both number and design. The problem was to link these permanently to a set of evolutions in a way that was clear, unambiguous and, as with any language, enjoyed a wide consensus. Admirals and commanders-in-chief had been expected to introduce their own systems. This might have worked well when squadrons exercised together and gained the confidence of an able commander, but a serious flaw that Ben Wilson highlights in *Empire of the Deep*[3] when a new C-in-C arrives or when two unfamiliar squadrons fight an action together. On the following pages we examine two battles of the American war where mixed messages adversely affected the outcome.

There was no single 'author' of the signal books that eventually wrested signalling from the dense text of the *Fighting Instructions* and provided a systematic tabulation matching signal to manoeuvre, though Admiral Lord Howe must take much of the credit for consolidating the work of Knowles, Kempenfelt, Rodney, Hood and others in his *Signal Book for the Ships of War of 1793* (see Timeline page 12).

If the evolution of the *Signal Book* was the fruit of experiment and experience in battle, the tactical manoeuvres contained within it were the subject of some debate and even rancour. One such was the signal (No 235) 'to break the enemy's line', the idea for which was claimed by an Edinburgh merchant and writer on naval affairs, John Clerk of Eldin. Due acknowledgement accorded to his tactical theories did not stop him seeking financial reward in a pamphlet of 1806.[4]

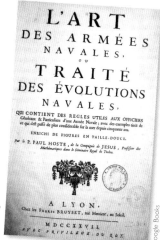

Title page (below) from 1727 edition of Père Paul Hoste's 1697 treatise on naval tactics translated into English in 1763. While not directly acknowledged, it was almost certainly an influence on Lord Howe in the development of his *Signal Book* of 1782 and is the first theoretical exponent of the tactic of breaking the enemy's line.

Google Books

Engraving after painting by James Saxon

John Clerk of Eldin also claimed originality for the tactic of breaking the line adopted by Rodney at the Battle of the Saintes in 1782 and thus the inspiration for subsequent naval victories.

41

Battle of Martinique
17th April 1780

In correspondence with the Earl of Sandwich, First Lord of the Admiralty, following the Battle of Martinique, Admiral Sir George Rodney laid the blame for his failure squarely on the 'dastardly behaviour of the fleet which calls themselves British...'.[5] This in itself reveals something of the reason of the inconclusive outcome, notwithstanding Rodney's claim in his formal post-action report that the enemy 'might be said to be completely beat'.[6]

The newly-arrived admiral's autocratic style of leadership as C-in-C Leeward Islands had not gained the confidence of those who served under him, on whom he later heaped considerable abuse.[7]

The battle itself was fought off the west coast of Martinique with the two fleets, some twelve miles apart at daybreak, slowly converging on opposing courses in light winds. Rodney had the weather gage and at 0645 signalled to the fleet, in line ahead on the starboard tack, his intention to attack the enemy's centre and rear with his whole force. This signal was acknowledged by all his ships, but when, more than five hours later, he ordered them to bear down in line abreast and engage their opposites in the line, as he put it in his official report 'agreeable to Article 21 of the Additional Fighting Instructions', he intended his original order to stand. But whether to attack the rear, now the French van, some 4.8 miles ahead, or in accordance with standing instructions to contain the vanguard so that it could not turn and double the British line, Captain Carkett in *Stirling Castle* led the British van northwards, dissipating the

attack and denying Rodney the victory he had hoped for and that would not now happen until the Battle of the Saintes almost exactly two years later.

❶ 0830, Rodney orders line of battle abreast downwind to engage the French fleet.

❷ De Guichen wears onto near-parallel course with wind abaft the beam.

❸ Rodney orders fleet to haul their wind and restore line ahead, now on larboard (port) tack.

❹ 1010 fleet ordered to wear together and resume starboard tack.

At 1100 Rodney flies signal to prepare for battle and at 1150 **❺** makes the signal to bear down together and steer for their opposites in the enemy line. In line with his original instructions some five hours earlier, this was intended to overwhelm the 14 ships of de Guichen's centre and rear with all 20 of his own fleet before the enemy van could tack or wear again and return to their aid.

Van de Jadeu

Van de Jadeu

Actual track of British van

Original Line of advance of British fleet

Admiral's intended engagement of centre and rear.

Centre de Guichen

Van Hyde Parker

Wind

Centre Rodney

Rear de Grasse

Rear Rowley

Original Line of advance of French fleet

0   1   2   3   4 nautical miles

Plot showing relative position of French and British fleets at 1150. Line lengths are shown to scale.

42

Battle of Chesapeake Bay
5th September 1781

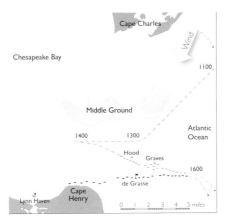

There can be few battles of relatively minor tactical outcome that had such a strategic impact as that fought off the entrance to Chesapeake Bay on 5th September 1781. Rear Admirals Sir Thomas Graves and Sir Samuel Hood's inconclusive action followed by several days of manoeuvre and counter manoeuvre off-shore allowed the landing of critical reinforcements that decided the land battle for Yorktown and the defeat of General Cornwallis. It was the end for Britain's North American colonies and as such led to inevitable recriminations at home. Questions were asked and pamphlets written pointing fingers of blame at both Graves as C-in-C and Hood, his second-in-command.

Aboard Hood's flagship *Barfleur* was a young midshipman Thomas White who, serving as a captain in 1830, wrote a carefully argued refutation of a damning critique[8] published some four decades after the battle of both admirals: Hood for failing to engage the enemy and Graves for not doing more to assist Cornwallis. His testament, among others, is valuable in understanding what happened on that afternoon and the critical part that signalling played.[9]

Attempts to construct an accurate narrative plot of events that unfolded over several hours are always foreshadowed by earlier versions, many re-cycled from uncertain sources which rarely agree on detail. For our purposes, an understanding of the signals made or not made is more illuminating than the precise dispositions of the two fleets whose relative positions are shown in the diagram above. De Grasse was

anchored off Lynn Haven with 24 ships-of-the-line when his scouting frigates warned of the approach of Graves's fleet of 19 ships. Unable to weather Cape Henry against the flood tide he had to wait until the ebb began at noon before slipping and forming a long line-of-battle close hauled on an easterly course.

An hour later, once clear of the Middle Ground bank, Graves signalled to his fleet to form a line-of-battle on a parallel but opposing course three miles to the north. He passed up the opportunity to attack the straggling centre and rear of the French fleet still tacking to clear Cape Henry, instead wearing his fleet together which reversed his order of battle. At 1430 he ordered his van to bear down toward the enemy, repeating the signal twice, making an even steeper line of approach. With the signal for the line, the Union

Admiral Lord Thomas Graves by Thomas Gainsborough, 1786.

Vice Admiral Sir Samuel Hood by James Northcote, 1784. A year older than Graves, their relationship was professional but not cordial. They had not met before their squadrons joined to confront de Grasse.

Below: A modern-day interpretation of the Battle of Chesapeake Bay, painted in 1962 by V. Zweg for the US Naval History Command.

Flag at the mizzen peak still flying, at 1546 he ordered the fleet to close to 1 cable (200 yards) followed by signals to bear down and for close action.

It is here that the source of the later arguments lie. Convention had it that the signal for the line-of-battle took precedence over any other signal and this was Rear Admiral Hood's defence when challenged as to why his rear division did not engage in support of the van and centre – irrespective of the fact that, with the steep angle of approach, his seven ships were still out of range, as were many of Graves's own division.

In countering the public criticism of 'the dilatoriness on the part of the rear division in obeying [the admiral's] signals and closing with the enemy'[10] Captain White was quick to point out that as  soon as the Union flag was hauled down at 'half past 5 pm... Sir Samuel Hood made the signal to bear up [the helm] and steer for the enemy's rear'.[11] But within ten minutes the signal for the line was re-hoisted on the flagship and Hood's ships ordered by pennant number to re-take their station in the line. To further emphasise his certainty that the signal for the line had been kept flying throughout the approach and eventual engagement, White goes on to say that if it wasn't flying '...what I that day saw and heard was a mere chimera of the brain and that what I believed to be the signal for the line was not a union jack but an *ignis fatuus* [will o' the wisp] conjured up to mock me'.[12] Fifty years after the battle, his passionate defence of Admiral Hood's adherence to the *Fighting Instructions* was impressive, though we can never know with certainty the precise sequence of the conflicting signals made that day.

Vice Admiral Sir George Tryon (1832-1893) a larger-than-life character beloved by the public for his dashing achievements in annual naval manoueuvres but regarded with suspicion by many contemporaries in the Service for his unorthodox methods.

HMS *Victoria* (right) that would become Tryon's flagship as C-in-C Mediterranean, seen here a year after her launch from the Tyneside yard of Armstrong Mitchell in 1887. Her low freeboard forward to accommodate her massive 16.25in guns (not yet fitted in this photo) earned her and her sister ship *Sans Pareil* the soubriquet 'A pair of slippers'.

Loss of HMS *Victoria*
22nd June 1893

There is a terrible irony in the events that unfolded in the mid-summer heat off the Syrian coast on 22nd June 1893. There was no ambiguity of signal and nothing exceptional in the routine though elegant choreography of the C-in-C's proposed manoeuvre to bring the two divisions of his Mediterranean fleet to anchor off Tripoli.

To appreciate the extent of the tragedy both for the Royal Navy and for the public at large, who held Sir George Tryon in high esteem, we have to remember the long iterative process of consultation and refinement that the *Signal Book* and signalling practices had undergone in the century since the actions of the American War. One of the principal protagonists in the polarised contest between the initiative-stifling control of the signal book tacticians and the encouragement of subordinates to think strategically was George Tryon. Richard Hough in his 1959 *Admirals in Collision* identified the damage done to the Victorian navy who had 'forgotten the fog of battle' due, in Tryon's view, to an over dependence on the 'thoroughly dangerous cult of signalling'.[13]

Tryon enthusiastically championed initiative and opportunity which did not sit easily with many of his

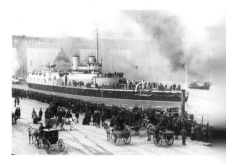

contemporaries. He contended that signalling by the book would fail in battle and that if everyone knew to take a lead from the flagship there was no need for signalling. Against stiff opposition, Tryon proposed a system for manoeuvring based on the two-flag hoist TA. Meaning 'Observe very attentively the admiral's motions...', TA had long been in the signal book with emphasis on maintaining station under sail once darkness fell; it now took on a new role as shorthand for his system. On assuming command in the Mediterranean in September 1891, he wasted no time in exercising his squadrons in the TA system, asserting in his first report to the Admiralty in November that the fleet 'manoeuvred with rapidity and accuracy on the plan to which these papers refer...'.[14]

The response in correspondence from fellow flag officers and his captains

With no signal to instruct columns to turn in succession inwards, separate hoists were required to address each division. Above left: Division 1 to turn in succession 16 points to port; above right: Division 2 to turn in succession 16 points to starboard. Only when all ships had repeated the signal close-up could it be made executive. Below: plot of intended manoeuvre with 2-cable safety margin.

was mixed but generally favourable, though some were more cautious. Among the latter was Rear Admiral Sir Albert Markham who joined Tryon as his second-in-command aboard HMS *Camperdown* in March 1892. Markham did not disguise his misgivings over the TA system, directly opposing Tryon in doubting that it would work in action. While the tragic events of 22nd June during the manoeuvres of the following year had nothing to do with the TA system, historians differ on the contrast in personalities between Tryon and Markham that may or may not have contributed to the fateful outcome. What is not in doubt is what happened, though we can never know what Tyron's true intentions may have been.

The plot on the left shows the intended manoeuvre to bring the two divisions to anchor which the Admiral had discussed with his Flag Captain and Flag Lieutenant. With a turning circle, or tactical diameter, of 4 cables and allowing a safety margin between the two columns, the required distance between the two columns should have been ten cables; but Tryon signalled the turn with the columns only six cables apart. Concerned at the risk of collision, Markham, leading the 2nd Division, delayed hoisting the repeat signal in acknowledgment, only to be stung in a public rebuke by semaphore from Tryon: 'What are you waiting for?'. The signal was hauled close-up and the order executed with inevitable consequences. Had Tryon done it deliberately to test the initiative of his second-in-command with a last-minute alternative up his sleeve or was it a catastrophic error? Tragically, for Tryon went down with the *Victoria* and 357 others, the one testimony not available to the subsequent court martial was that of the Vice Admiral himself.

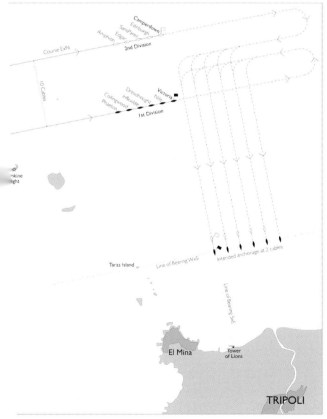

Reaching the Limit: Signal success and failure at Jutland

David Morris

Lapel badge celebrating Buntings Day.

Admiral Jellicoe's flagship HMS *Iron Duke* in the van of the 3rd Division 'Crossing the T' of the High Seas Fleet. Although apparently depicted after executing the deployment manoeuvre, he is still flying the equal speed signal repeated by HMS *Royal Oak* immediately astern. This may be artistic licence celebrating the signal.

On 31st May every year at precisely 1815 former Royal Navy 'buntings', as signalmen were known, raise a glass to the famous flag hoist made by Admiral John Jellicoe when he deployed his battle fleet in line ahead at the Battle of Jutland. The day is designated Buntings Day.

The hoist, comprising the signal flags C and L preceded by the Equal Speed pennant, was a precise instruction that read as follows: 'The column nearest SE x E is to alter course in succession to that point of the compass, remaining columns altering course leading ships together the rest in succession so as to form astern of that column, maintaining the speed of the fleet.' In the phonetic alphabet of the time the signal, also sent by W/T, was known as 'Equal Speed Charlie London' and signalled the moment at which Jellicoe decided

to form his battle line on his port column so as to 'cross the T' of the German High Seas fleet approaching from the SSW. It was to be the last time a manoeuvre on such a scale was executed in battle and, with the signal's adoption as a motif by the Royal Navy Signals School, a lasting testament to the effectiveness of flag signalling in the manoeuvring of large fleets.

The six divisions of battle fleet form in line ahead.

It is no surprise that out of more than 1,400 signal exchanges among ships of the Grand Fleet and Vice Admiral David Beatty's Battle Cruiser Fleet between

Signal Method	Combined Fleets		Battle Fleet to 1800*		Battle Cruiser Fleet to 1800*	
Flags only	719	(51%)	319	(55%)	234	(58%)
Searchlight only	323	(23%)	161	(28%)	83	(20%)
Masthead Morse lamp	125	(9%)	28	(5%)	22	(5%)
W/T only	128	(9%)	24	(4%)	39	(9%)
Semaphore only	85	(6%)	35	(6%)	26	(6%)
Combined methods	39	(2%)	8	(2%)	7	(2%)

*Time by which both fleets were effectively operating together.

Tabulation of signals from Appendix 2 of The Official Despatches (HMSO) broken down by method and between the main Battle Fleet and the Battle Cruiser Fleet up to 1800. Despite his disdain for the Signal Book, Beatty's BCF made a higher proportion of flag signals than his Commander-in-Chief.

midnight on 30th and midnight on 31st May, more than half were made by flag only, with daylight Morse signalling by searchlight accounting for just under a quarter of the total. The breakdown by method of the 1,419 signals made in that 24-hour period is tabulated above.

Signalling by light and semaphore played their part at Jutland and with these methods, covered later in the book, the record reveals a more spontaneous and human response to unfolding events. Our focus here is on the role of flag signalling, more than half of which in both fleets controlled speed and anti-submarine zig-zag routines in the long approach passage across the North Sea. To ensure the timely dissemination and execution of instructions, fast light cruisers, HMS *Boadicea*, HMS *Active*, HMS *Blanche* and HMS *Bellona,* were deployed with the main battle fleet as visual signal repeaters. HMS *Tiger,* at the rear of the Beatty's 1st Battle Cruiser Squadron, carried out the same repeating task to the two ships of the 2nd BCS and the four *Queen Elizabeth* class battleships of the 5th Battle Squadron. Grand Fleet Battle Orders also stipulated '...in the presence of the enemy all signals are to be made by flag, searchlight and wireless'.[1] However, once battle was joined, procedural failures led to lasting controversies, discussed on the following pages and which historians still debate.

My purpose here is not to rehearse the arguments, apportion blame or to pose counter factual scenarios in a set of circumstances that tested visual signalling to the limit. Nevertheless, despite the successful outcome of Jellicoe's deployment signal, it is important to understand what those limits were and the cultural context of what Andrew Gordon has called the 'Long Calm Lee of Trafalgar' in which this real-world test of signalling practices and reliance on the *Signal Book* took place.[2] As we have seen, long-gone were the confident initiatives of Nelson's 'Band of Brothers', subsumed by decades of rational improvement as sail gave way to the precision of 'steam tactics' in a Navy increasingly bound by order and unquestioning obedience.

Viewing Distance (yds)	1,000	2,000	4,000	6,000	8,000	10,000
Minutes of arc	10.20'	5.10'	2.58'	1.74'	1.26'	1.02'
Height to naked eye*	1.48mm	.74mm	.37mm	.25mm	.18mm	.15mm

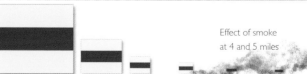

Apparent size with telescope (15x)

Effect of smoke at 4 and 5 miles

*equivalent image size at comfortable reading distance

Left: Comparative visibility of Size 1 signal flags (9ft × 11ft) under optimum conditions at half a mile (1,000yds) up to five miles. For comfortable legibility to the naked eye, the visual angle should be at least 20 minutes of arc. Add in funnel smoke, gunfire and mist and the challenges are obvious.

Turn to the South East
1415–1535

Vice Admiral Sir David Beatty.

Flag hoist Pennant 9 DH: Alter course lead ships together, rest in succession, SSE.

Plot of the 5th Battle Squadron and Battle Cruiser Fleet based on combination of British Admiralty post-action plots prepared under the supervision of Captain J E T Harper RN and those for the German High Seas Fleet prepared by Captain Otto Groos (*Der Kriege in der Nordsee*, 1920).

Had Evan-Thomas been given access to Beatty's Battle Cruiser Fleet Orders, he would have known that the sharp alteration of course at 1415 was a likely indicator of imminent action (see Gordon, p. 88).

A t 0122 on the morning of 31st May Admiral Jellicoe signalled the Admiralty with his proposed 1400 rendezvous positions for the Grand Fleet and Vice Admiral Beatty's Battle Cruiser Fleet with the 5th Battle Squadron. Not quite 13 hours later and 10 miles NW of his designated position at 1415 ❶, Beatty made the first of a series of course alterations that broke the mould of the routine zig-zag passage, turning his fleet sharply North in the hope of making contact with the advance cruisers of Jellicoe's fleet. He had just received a signal by searchlight from HMS *Galatea*, at a range of 13 miles, reporting what would turn out to be a light cruiser of Admiral Hipper's 2nd Scouting Group. What happened over the next 80 minutes has been forensically examined many times, but anomalies still emerge which bear further examination.

Starting at 0230 in the long pre-dawn twilight of the northern summer, the routine signals controlling zig-zag patterns had been made by flag signal from Beatty's flagship HMS *Lion*, repeated by searchlight by HMS *Tiger* at the rear of his column. The signal record shows that these were mirrored almost exactly by flag signals from HMS *Barham*, Rear Admiral Hugh Evan-Thomas's flagship, to the ships of his 5th Battle Squadron, stationed some five miles on Beatty's port quarter. The routine was working well until Beatty's 1415 signal to alter course to N x E which his Flag Lieutenant, Lt. Commander Ralph Seymour, made by searchlight even though they were well within flag signalling distance. Two minutes later the 5th BS made their turn, the signal repeated to his column by flag. But the next critical alteration of course, some 145° to starboard to SSE signalled at 1432 ❷ with the flag hoist 9DH was missed. By that time *Barham* was only 4.2 miles on *Lion*'s port bow but the signal from the flagship

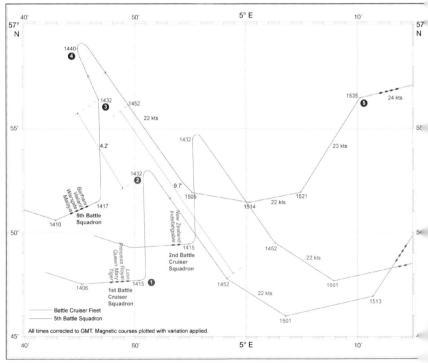

was not repeated by HMS *Tiger*, an omission that has been hotly debated. As the two columns of the BCF began their turn to starboard, Evan-Thomas made a two-point turn to port as if continuing with the zig-zag pattern; they were thus on a near-reciprocal course diverging at a combined speed of 42 knots. It is worth noting that in the official record the 5th BS flag signal to alter two points to port is recorded *before* Beatty's signal. This may be of no significance, with the inevitable variation in time keeping between ships and times rounded to the nearest minute, but could suggest that, though both signals are timed at 1432, Evan-Thomas had already decided to resume his zig-zag pattern ❸. Whatever the case, a further eight minutes passed before he took the decision to turn in pursuit of the BCF and made the signal to his squadron to alter course in succession to the SSE, increasing speed to 22kts ❹. By then he was nine miles astern of Beatty's column and during that interval, there is no record of any attempt by the flagship or any of the battle cruisers to communicate with the 5th Battle Squadron by light signal.

As the 5th BS was making its turn to the SSE, Beatty had received a further sighting report from *Galatea* by W/T of 'large amounts of smoke as though from a fleet' to the ENE but continued SSE for a further 12 minutes before altering to SE by which time Evan-Thomas was 9.8 miles astern. Subsequent alterations to E and NE allowed the 5th BS to close the distance so that their alteration of course to NE at 1521 brought the BCF onto their starboard beam at a distance of 5.9 miles. At 1527 Beatty made a general signal by flag 'Assume complete readiness for action in every respect'; the exact same signal was repeated by Evan-Thomas, also by flag, to his own column three minutes later, suggesting that the two fleets were again within flag signalling range. The next five minutes saw a further flurry of signals by flag, searchlight and W/T clearly establishing the position of the Battle Cruisers of Hipper's 1st Scouting Group, including the first ever such spotting report from HMS *Engadine*'s seaplane. Then at 1535 a dramatic but ambiguously worded signal by searchlight from Beatty to 5th BS: 'Speed 25 kts. Assume complete readiness for action. Alter course lead ships together, rest in succession to E. Enemy in sight.' Ships already in line ahead, as the 5th BS were, can either alter course together or in succession, there being only one lead ship. Evan-Thomas simultaneously signalled to his squadron by flag: 'Alter course in succession E, speed 24 kts' ❺. He had half a knot in hand before reaching the maximum designed speed for his *Queen Elizabeth* class battleships.

Ten minutes later, now racing east at 25 kts, Beatty deployed his six Batttle Cruisers, now in line ahead, on a line of bearing NW to allow all to clear their smoke and bring their forward turrets to bear on Hipper's 1st Scouting Group. At 1547 he made the single flag hoist No 5 'Open fire and engage the enemy', at a range of 18,500 yards. A further 19 minutes would pass before Evan-Thomas was able to bring his heavier-calibre guns to bear on the rear of the German battle cruisers in support of the BCF. These were critical moments, that despite Beatty's warm words in his official Report of Proceedings: 'Led by Rear Admiral Evan-Thomas MVO, in *Barham* [the Fifth Battle Squadron] supported us brilliantly and effectively',[3] were to lead to deep and lasting recriminations.

Flag hoist BJ: Assume complete readiness for action in every respect.

Numeral 5: Open fire and engage the enemy, a signal to be executed as soon as it was seen. Eight minutes later Beatty signals with hoist 65 repeated by W/T: Increase rate of fire (below).

Turn to the North
1630-1710

Two documents, or rather the lack of one of them, among many issued after the battle testify to the divisions that characterised the debate over the conduct of the fleets at Jutland. The first is an amendment made by Vice Admiral David Beatty to his Battle Cruiser Orders that includes the following sentence:

'It therefore becomes the duty of subordinate leaders to anticipate the executive orders and to act in the spirit of the Commander-in-Chief's requirements.'

This is a clear reiteration of Beatty's standing Fleet Orders, to which Rear Admiral Hugh Evan-Thomas had never been given access and which urged each captain '...to use his discretion in handling his ship as he considers that the Admiral would wish.' The other document – the fair copy signal log from Beatty's flagship, is mysteriously missing half a page that should have recorded the outgoing signals following the flag hoist ordering the 5th BS to turn the north. The significance of both will become clear by following the sequence of events as the Battle Cruiser Fleet closed the main battle squadrons of the High Seas Fleet fast approaching from the south.

When HMS *Southampton,* scouting with the 2nd Light Cruiser Squadron, signalled by searchlight the first sighting report of the High Seas Fleet at 1633, Beatty's BCF, with the 5th BS some 8 miles astern, had been heavily engaged with the battle cruisers of Hipper's 1st Scouting Group for 46 minutes, suffering the loss of two of his six ships: HMS *Indefatigable* and HMS *Queen Mary.* At 1638 Beatty received a more detailed sighting report from *Southampton* by W/T and two minutes later ordered a general signal 'Alter course in succession 16pts to starboard' made by flag only. ❶ Given the difficulty in correctly reading the SSE turn signal

Flag hoist CT: Alter course in succession 16 pts to starboard.

Plot of the 5th Battle Squadron and Battle Cruiser Fleet with Scheer's 1st Scouting Group and elements of Hipper's High Seas Fleet during the 'Turn to the North' (Sources as plot on page 48).

Note that light cruisers scouting ahead and destroyer flotillas who were heavily engaged between the opposing Battle Cruiser fleets are omitted for clarity.

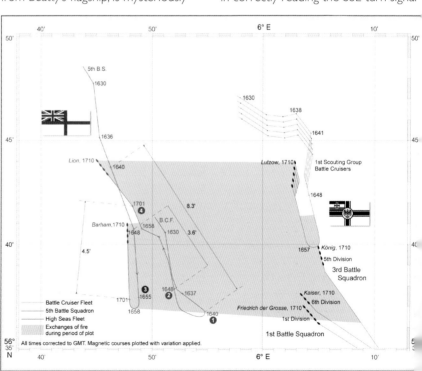

at 1432 at a distance of less than five miles, it seems unlikely that the 'general' address was intended to include the 5th BS, then 8.3 miles to the NW.

Having made his turn, Beatty now had the 5th BS, still heading SSE, 2 pts on his starboard bow. He could have continued to the NW to allow the two squadrons to approach each other safely starboard to starboard before ordering Evan-Thomas to execute a turn to starboard to fall in astern of him; instead he altered course a further 4 pts to the north to pass to the east, on the *engaged* side of the 5th BS. At 1648, with the two squadrons separated by only 3.6 miles and closing at a combined speed of 48 kts, Beatty signalled by flag specifically to 5th BS a repeat of his signal at 1640 ❷. But with *Lion* steaming at 25 kts and the light westerly breeze, the flags, among many signals for which Seymour was responsible at that critical time, would have been flying at such an angle as to be impossible to read from *Barham*'s flag bridge until the ships were almost abreast at approximately 1651. Only then would Evan-Thomas's signalmen have been able to acknowledge with the answering pennant at the dip before hauling it close-up to indicate that the signal had been understood. Standard procedure required that they would then wait for the order to be made 'executive' on being hauled down on the flagship, before making their turn. But, for whatever reason, and it must here be acknowledged that Seymour was under intense pressure, the signal was not made executive for at least three minutes, by which time Evan-Thomas's battleships were already three miles astern of Beatty's combined squadron and had come within range of the 12in guns of Admiral Scheer's 3rd Battle Squadron.

As Evan-Thomas began his turn at 1655 he made the flag hoist TA 'Observe attentively the Admiral's motions' ❸. The irony of that choice, the rallying cry of Vice Admiral Sir George Tryon, of whose tireless campaign to liberate the Navy from what he regarded as the tyranny of the *Signal Book* we have already heard, would not have been lost on his flag captain and staff. But under the concentrated fire of Scheer's *König* class dreadnoughts as his four ships were turning in succession silhouetted against the sunlit horizon to the west, it was not meant ironically. The signal meant what it said: to closely follow his lead without further signals as they reversed course and re-trained their main armament to the starboard quarter while adjusting course to stay in touch with the BCF now 4.5 miles ahead. At 1701 Beatty somewhat unnecessarily signalled 'Prolong the line by taking station astern' ❹ – a task made no easier by the faster battle cruisers pulling further ahead.

And the two documents? Beatty's amendment to his Fleet Orders, Gordon argues, could only be seen as a critique of what he perceived as Evan-Thomas's failure to act on his own initiative on two occasions during the afternoon. This polarisation of individual initiative versus a doctrine of central control lies at the heart of the Beatty/Jellicoe debate that has divided opinion for more than a century. As for the missing signal record, Andrew Gordon leaves hanging the question as to whether or not its omission could have been at Beatty's request to protect the questionable competence of his chosen flag lieutenant who later followed him to the Admiralty.[4] Sadly for Evan-Thomas the post-war struggle to vindicate his decisions took its inevitable toll on his health. Nobody came out of it well.

Flag hoist TA: Observe attentively the Admiral's motions.

Flag hoist 71: Prolong the line by taking station astern.

[PD-UKGov]

Rear Admiral Hugh Evan-Thomas who led the 5th Battle Squadron aboard his flagship HMS *Barham*.

If the Battle of Jutland pushed flag signalling to the limit, it was by no means the end of visual signalling in the face of growing use of W/T. Signalling by flag, flashing light and semaphore remained and still remains the only secure method for fleet communication with ships in sight of one another.

Lessons had been learned from Jutland and other actions during the war and in particular from the introduction of the convoy system in 1917. With war looming again in 1935 the Admiralty issued a substantial Naval Appendix to the International Code of Signals which detailed signalling protocols between ships in convoy and their naval escorts. Every possible convoy manoeuvre was covered in three, four and five-flag hoists including the one signal that no merchant ship master wanted to see addressed to his ship: ZGJ (left)– 'If you cannot maintain present speed of convoy you will have to proceed independently unescorted.'[1]

Two years later a revised Naval Code was issued (see page 37), designating seven flags: C, E, F, G, H, M and R from the 1934 International Code as special naval flags to meet changing needs – not least naval aviation, with flag 'F' still in use today to signal fixed-wing flying operations from an aircraft carrier.

By 1944, the Naval Code had grown to incorporate numeral and special flags from the US Navy and would, alongside the International Code, become the basis of the Allied Maritime Tactical Signalling and Manoeuvering Book (known as ATP-01, Vol II) with the formation of NATO in 1949. Regular updates of ATP-01 are issued by the NATO Standardisation Office to the present day. The NATO 'flag bag' now contained 68 flags: 26 alphabet flags and 10 numeral pennants from the International Code, 10 numeral flags from the US Navy, four substitutes (three from the International Code plus a fourth from the USN) and 18 special flags, most of which had been introduced from the US for combined fleet signalling in 1944 with some changes of meaning. Shown left is the page of numeral pennants and special flags from a 1982 copy of the US Navy Signalman Training Manual.[2]

[PD-UK-Unknown]

North Atlantic convoy during World War II with hoist ZGJ above.

US Naval Education and Training Command

PENNANT and NAME	Spoken	Written	PENNANT or FLAG	Spoken	Written	PENNANT or FLAG	Spoken	Written
PENNANT ONE / 1	PENNANT ONE	p1	CODE or ANSWER	CODE or ANSWER	CODE or ANS	NEGATIVE	NEGAT	NEGAT
PENNANT TWO / 2	PENNANT TWO	p2	SCREEN	SCREEN	SCREEN	PREPARATIVE	PREP	PREP
PENNANT THREE / 3 ·	PENNANT THREE	p3	CORPEN	CORPEN	CORPEN	PORT	PORT	PORT
PENNANT FOUR / 4	PENNANT FOUR	p4	DESIGNATION	DESIG	DESIG	SPEED	SPEED	SPEED
PENNANT FIVE / 5	PENNANT FIVE	p5	DIVISION	DIV	DIV	SQUADRON	SQUAD	SQUAD
PENNANT SIX / 6	PENNANT SIX	p6	EMERGENCY	EMERGENCY	EMERG	STARBOARD	STARBOARD	STBD
PENNANT SEVEN / 7	PENNANT SEVEN	p7	GROUP/ FLOTILLA	GROUP/ FLOTILLA	GROUP/ FLOTILLA	STATION	STATION	STATION
PENNANT EIGHT / 8	PENNANT EIGHT	p8	FORMATION	FORMATION	FORM	SUBDIVISION	SUBDIV	SUBDIV
PENNANT NINE / 9	PENNANT NINE	p9	INTERROGATIVE	INTER-ROGATIVE	INT	TURN	TURN	TURN
PENNANT ZERO / 0	PENNANT ZERO	p0						
TACK LINE	TACK	—						

SUBSTITUTES

1st. SUBSTITUTE	FIRST SUB	1st.	3rd. SUBSTITUTE	THIRD SUB	3rd.
2nd. SUBSTITUTE	SECOND SUB	2nd.	4th. SUBSTITUTE	FOURTH SUB	4th.

EXERCISE "MAINBRACE"

Although the Royal Navy had operated with US and Royal Canadian naval forces in the North Atlantic and with the Royal Australian and New Zealand Navies in the Mediterranean and the Pacific during World War II, the task of harmonising visual communications between the navies of the twelve founding members of the NATO alliance was formidable and urgent. The first major naval exercise undertaken by Allied Command Atlantic, codenamed 'Mainbrace', took place in September 1952 off Norway's North Cape; 203 ships representing nine nations took

part including 66 ships of the Royal Navy. Among them was Britain's newest and last battleship, HMS *Vanguard,* seen in the Associated-British Pathé's newsreel coverage of the exercise (reconstructed title still shown left).

Beside NATO's standardisation of signalling procedures, the UK, the United States, Australia, New Zealand and Canada also share responsibility for allied communications through the Combined Communications-Electronics Board. ACP130 (A) *Communications Instructions Signalling Procedures in the Visual Medium* sets out a 'concise and definite language'[3] for visual communications in all mediums: flags, flashing light, semaphore, pyrtoechnics and panel signalling, the latter included in the NATO wall chart of 2018 below. Visual signalling is alive and well.

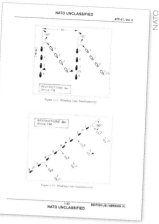

Page setting out squadron wheeling manoeuvres from ATP-01, Vol II; Bigot de Morogues (see p.11) would have approved.

NATO signal chart publicly available to download from www. nato.int/alphabet.

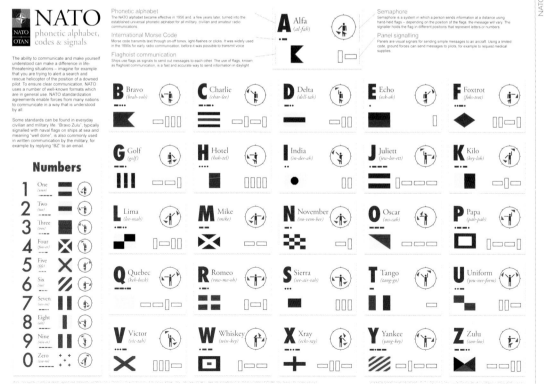

Bravo Zulu: Merchandise and marketing

The bold designs and colours of the International Code of Signals, optimised over a long process of iteration to achieve maximum clarity, are a ready-made palette of motifs and meanings for graphic designers. Shown on these pages are just some of many adaptations that have found their way onto mugs, tee-shirts, tote bags, signage and corporate branding.

The story of what was possibly the first such use is told by Captain Barrie Kent – an enamelled brooch of a flag hoist given by King Edward VII to his mistress Alice Keppel after sailing at Cowes Week. There was little ambiguity in the signal from the Code Book, Flag Z, Pennant 9, Red Burgee, Flag M: 'Position quarterly and open, I am about to fire a Whitehead torpedo ahead'.[1]

Less risqué and perhaps better known is the two-flag hoist adopted by NATO and used widely in the military to commend a task well executed: Bravo Zulu.

Right: Branding for Royal Docks based on code flags R and D plays on the motifs of 'plus' and 'equal'.

Burger bar branding, Washington DC.

Left: Branding for Royal William Yard Festival, Plymouth.

Below right: Brand mark for swim and beachwear, Spain based on letters B 'Bravo' and C 'Charlie'.

Studiose, Plymouth (www.studiose.co.uk)

1.1 Logo
Introduction

The international nautic code system (ICS)[1] consists of individual flag designs for each letter of the alphabet. COMMONWEALTH's three stripes logo is derived from the letter "C" — Charlie — which in nautic language means "Yes".

Left: Branding for New York agency 'Commonwealth' based on code flag C.

Below: Change of colourways used in branding for South Devon music festival.

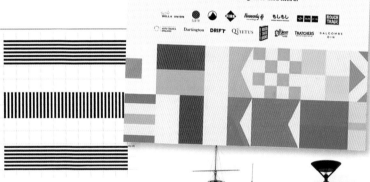

1.1 Logo
Introduction

COMMONWEALTH's three stripes logo represents the letter "C" derived from the international nautic code flag system combined with heraldic's black & white pattern based international color reference system. The flag's initial colors (blue/red) are replaced with its heraldic pattern code equivalent. (▬/▥)

3.1 Brand Type
Introduction

CW Flag™ Regular is the name of the agency's brand typeface. Simple letter shapes designed from basic graphical elements such as the square, circle and triangle form a new distinctive alphabet and a strong branding tool reminiscent of the graphic language of ensigns and ICS.

Signage at Portsmouth Historic Dockyard.

Starter's Orders: Signal flags in yacht racing

The popularity of signal flags as merchandise and marketing tools perhaps owes something to the continued use of International Code Flags in the management of yacht and dinghy racing. Sailing is not a spectator sport and the starting process may seem arcane, but while athletes crouch on their blocks awaiting the starter's gun, the start of a yacht race is not quite so straightforward, as anyone who has made the long sail out to the start line, often among dozens of other boats of different classes, will know.

Ever since the rules governing what were called sailing matches were agreed under the auspices of the Yacht Racing Association in 1875, the format has been broadly the same – a warning signal, a preparatory signal and a start signal. What has changed over time is the intervals between these three and the addition of several flags to indicate events such as postponement, a recall or that particular rules and penalties will be applied. Many of the original rules will be familiar to competitors today – though perhaps not Rule 14 'There shall be no limit as to the number of paid hands... [but] no paid hands may join or leave a yacht after the signal to start'![1] With large fully-crewed yachts lining up in tidal waters, many starts were made from moored buoys, with sails only fully set once the start gun had gone and a warning period of thirty minutes. The alternative, a running start, we are more familiar with today, with a start line at right angles to the course to the first mark, usually directly up-wind. The flags and present day timings are shown here.

Class warning flags from 1879 racing rules. The flags 'B', 'C', 'D' and 'F' are from the 1857 Commercial Code (see page 30) denoting races 1,2,3 and 4. Classes and time allowances were determined by Thames Tonnage.

5

Start -5 minutes. Gun.

Warning signal, class flag displayed.

4

Start -4 minutes. Gun.

Preparatory signal, 'Blue Peter' with class flag. Additional flags 'I' and 'Z' may warn of penalty rules in effect.

I

Start -1 minute. Gun.

'Blue Peter' lowered leaving class flag only.

0

Start. Gun.

Class flag lowered (to dip if boats over the line until all have returned to re-start). Next sequence starts.

Aide-memoire for racing flags and their meanings.

Close sailing at start of 2009 Laser World Championships, Halifax Nova Scotia.

UNIVERSAL TELEGRAPH.

TABLE *of the* SIGNS *or* COMBINATIONS.

Positions	Appearance		Positions	Appearance	
	By Day	By Night		By Day	By Night
1			25		
2			26		
3			27		
4			34		
5			35		
6			36		
7			37		
12			45		
13			46		
14			47		
15			56		
16			57		
17			67		
23			STOP		
24			FINISH		

Reconstruction of Plate 2 from Charles Pasley's 1823 *Description of the Universal Telegraph for Day and Night Signals.*

Timeline: Shapes, shutters and semaphore

If the development of flag signalling was carried out at sea, the process was very different for semaphore, which was not adapted for use at sea until the second decade of the nineteenth century and only then experimentally.

Semaphore has its roots in several experiments in 'distant writing' – the literal translation of the Greek words *tele* (distant or far) and *grapho* (writing) and many of the devices were known as 'telegraphs', often misnamed as 'semaphores' and vice versa. Both are effectively distance signals animated either by shutters or articulated arms. It is the latter, still in use today with hand flags replacing steel or louvered wooden arms, which is generally termed semaphore – literally translated from Greek as carrier (*phero*, to bear) of signs (*sema*). To understand how we got here, the timeline on these pages traces the colourful history of the devices and their various champions among whom were eminent scientists, a French Abbé, an 18th-century rake, two English clergymen, a rear admiral and a lieutenant colonel of the Royal Engineers*.

* As with the work of W G Perrin on signal flags, any fresh look at the history of semaphore must acknowledge the scholarship in the 1930s of Commander Hilary Mead RN. See endnotes.

1684	Notwithstanding the listing of 'A Mute and Perfect Discourse by Colours' which describes 'How, at a window, as far as eye can discover black from white, a man may hold discourse with his correspondent' in the Marquis of Worcester's 1673 *Century of Inventions*,[1] it was the eminent scientist and polymath **Sir Robert Hooke** who can claim to be the first to have proposed an organised means of telegraphing information over long distances. In a paper delivered to the Royal Society in 1684, Hooke describes a system of shapes representing letters of the alphabet hung within a wooden frame that could be repeated from station to station a fixed distance apart, such that 'none but the two extreme correspondents shall be able to discover the information conveyed'.[2]
690	**Guillaume Amontons**, a French physicist and instrument maker, who had earlier presented a system similar to Hooke's symbolic telegraph, demonstrates a new telegraphic device to the Dauphin in the Jardin du Luxembourg.

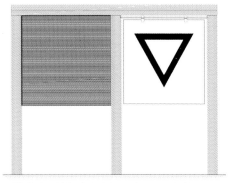

Sir Robert Hooke's telegraph presented in 1684. Successive shapes were to be pulled out from the housing to spell out words in his own coded alphabet. The example below spells out 'HOOKE'.

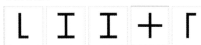

A contemporary illustration shows a single moveable arm on a short flagpole but few other details; it pre-dates by just over 100 years the first true semaphore.

1767   **Richard Lovell Edgeworth**, an Anglo-Irish inventor, politician and a member of the Birmingham Lunar Society, sets up a relay of signal apparatus with stations 16 miles apart between London and Newmarket. Funded by his eccentric friend Sir Francis Blake Delaval who was anxious for news of the horse racing,[3] this private line was probably the first successful telegraph and may have used a forerunner of the device that he would later advocate to the Royal Irish Academy in 1795.

Sir Francis Blake Delaval 1727-1771, after Joshua Reynolds. Edgworth's telegraph from Newmarket ended on the roof of Delaval's townhouse, No 11 Downing Street.

Image: Bonhams

1792   **Abbé Claude Chappe**, one of whose four brothers, Ignace, was a member of the Legislative Assembly, successfully demonstrates his 'T' telegraph in Paris. Its articulated arms are capable of 196 different positions but only 92 most easily distinguished combinations are used. In the previous year, against a background of revolutionary unrest, he had demonstrated a shutter system and a 'synchronous' telegraph in which the hands of a dial indicated symbols similar to those used by Hooke, only to have the equipment for both of these systems seized and burnt by the mob fearing they were being used to communicate with the imprisoned King Louis XVI.[4]

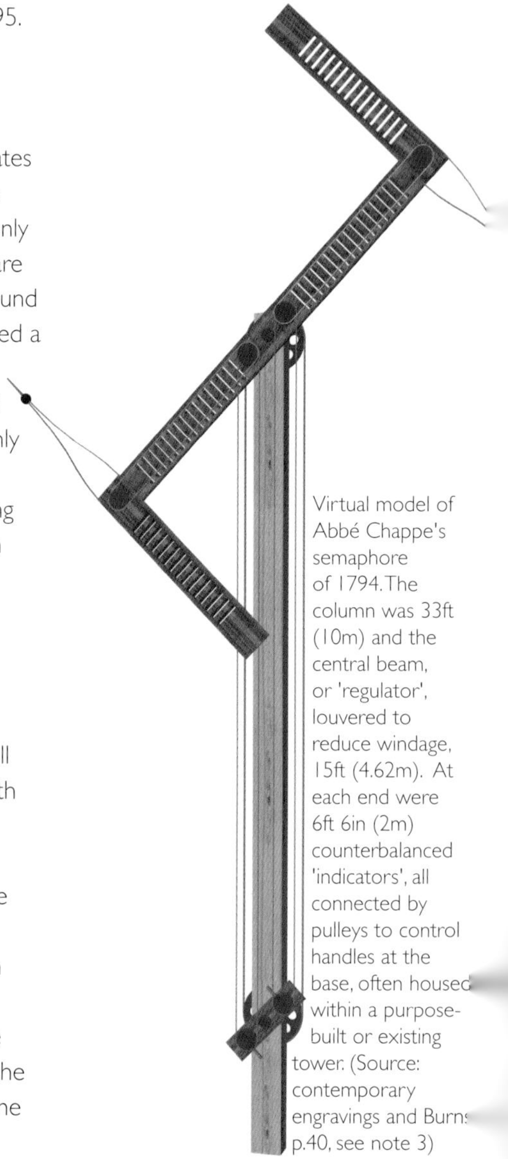

Virtual model of Abbé Chappe's semaphore of 1794. The column was 33ft (10m) and the central beam, or 'regulator', louvered to reduce windage, 15ft (4.62m). At each end were 6ft 6in (2m) counterbalanced 'indicators', all connected by pulleys to control handles at the base, often housed within a purpose-built or existing tower. (Source: contemporary engravings and Burns p.40, see note 3)

1793   Despite a similar setback with the firing of his first experimental semaphore, Chappe secured the support of the Assembly, now re-named the National Convention, for a full trial. This took place two days before the fifth anniversary of the storming of the Bastille and led Joseph Lakanal, one of the three Convention members appointed to oversee the trials, to wax lyrical in his somewhat chauvinistic praise of Chappe's invention on 26th July 1793:

> 'What brilliant destiny do science and the arts not reserve for a republic which, by the genius of its inhabitants, is called to become the nation to instruct Europe.'[5]

1794

Chappe's system was approved by the Convention and by August 1794 a line between Montmartre and Lille was established. On 15th August the first transmission brought the news in under two minutes that the border town of Le Quesnoy had been re-taken from the Austrians with whom, and her allies Britain and Holland, France had been at war since April 1793. Driven by the imperative and proven benefit of good communications in wartime, it was the first of a rapidly expanding network of semaphore chains that by 1815 covered 1,780km (1,112 miles) with 224 stations. By the time the electric telegraph arrived in 1844, the semaphore network extended to 534 stations connecting 29 major towns and cities in France, with extensions into Belgium, Holland and Milan. Tragically, Chappe did not live to see the full effect of his invention – dogged by ill health and undermined by rival claims to the design of his system, he took his own life in January 1805.

PARIS. — MONUMENT CLAUDE CHAPPE
1er septembre 1944, 150e anniversaire de la prise de Condé, annoncée par le télégraphe aérien de Chappe à la Convention nationale. Celle-ci adresse le même jour à l'armée un télégramme de félicitations.

Postcard view of Chappe's monument in Paris and commemorative stamp issued in September 1944 to mark the 150th anniversary of the first transmission.

1795

Almost exactly a year after the successful launch of Chappe's semaphore in Paris and nearly thirty years since his Newmarket experiments, **Richard Edgeworth** assisted by his son (probably his third son Lovell, the seventh of his twenty-two children by four wives) demonstrated a new version of his telegraph, the first in the British Isles to employ rotating arms. Using an indexed code based on multiples of seven possible positions of the triangular pointers – the eighth, pointing upwards, was the rest position – they successfully transmitted four messages across the North Channel between Donaghadee and Portpatrick, a distance of twenty miles. But, despite early encouragement from the Admiralty, the isosceles triangles were judged to be too indistinct and the decoding process too cumbersome to make it viable and, to Edgeworth's disgust and deep frustration, his scheme was turned down. A pamphlet of 1797 defending his 'Tellograph', aimed at the Earl of Charlemont, President of the Royal Irish Academy, did not exactly help his cause.[6]

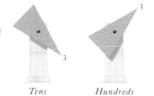

Units *Tens* *Hundreds* *Thousands*

Edgeworth's telegraph re-drawn from engraving in Rees's *Cyclopedia*, 1819 (see note 6). Rotating clockwise from viewers position in 45° increments, the arms indicated numbers 0 to 7.

At the same time that Richard Edgeworth was demonstrating his 'Tellograph' to the Irish Academy in August 1795, the first of two new telegraph systems, almost identical in principle, was being demonstrated to Admiral Sir Peter Parker, C-in-C Portsmouth on Portsdown Hill.

News of Chappe's success had reached England soon after the first successful transmissions, though sources differ on the route by which it came. Probably the most authentic is the account given by the Reverend **John Gamble**, Chaplain to the Duke of York, commander of the British Army in Flanders. In his *Essay on the Different Modes of Communication by Signals* published in 1797 and fulsomely dedicated to the Duke, he tells of a French prisoner in whose pocket a drawing and description of Chappe's semaphore was found and, following a detail description of the apparatus, he relates how he suggested to the Duke that '...a machine might be constructed to answer the purpose better.'[7]

Gamble was careful to acknowledge the work of both Chappe and Edgeworth, and one or two more eccentric proposals, before outlining how he had brought his own proposals for a shutter based telegraph to the Admiralty from whom he received an order on 19th July for a trial between Portsdown, Spithead and the Isle of Wight, some fourteen miles away. Despite the apparent success of the trial witnessed by Admiral Parker on 6th August, he learned three weeks later 'in a desultory conversation with one of their Lordships'[8] that a rival shutter system proposed by the Reverend **Lord George Murray**, second son of the 3rd Duke of Athol, was preferred. Was it hubris at French advantage with Chappe's semaphore that the shutter system was adopted in England? Was Murray, whose system the Admiralty assured Gamble they had seen 'above a twelvemonth before', better connected? Or was his six-shutter telegraph, which Gamble likened to '.. adding a fifth wheel to a carriage which, on many accounts, would be inconvenient',[9] just that much better than Gamble's four shutters?

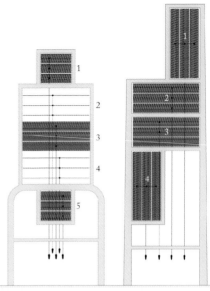

Gamble's five and four-shutter telegraphs based on engraved plates from his 1797 *Essay of Different Modes of Communication by Signals*. The horizontal shutters were 7ft ×16ft (2.13m × 4.88m); it was the four-shutter system erected between spare topmasts of two 74s that Gamble demonstrated to Admiral Parker.

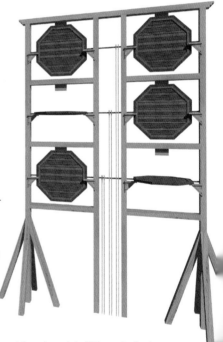

Virtual model of Murray's six-shutter telegraph offering 63 different permutations. Each octagonal shutter was 5ft (1.5m) square and the whole structure stood 28ft 6in (8.7m) tall. (Source: Broadsheet published by S.W. Fores of Piccadilly, 25th March 1826)

Despite his disappointment at the Admiralty's preference for the Murray shutter system, the Rev. Gamble soon switched his attention to a new proposal for a five-arm 'Radiated Telegraph'. It was the first moving-arm telegraph proposed in Britain.

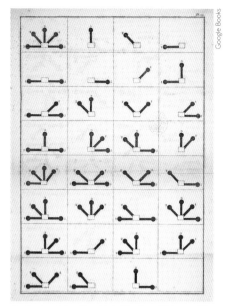

1796 Having made their decision, the Admiralty wasted no time in commissioning Murray, later elevated to a new post created by King George III in March 1796 as Director of Telegraphs to the Admiralty, to build a chain of fifteen telegraphs connecting London and Deal. On 27th January, the first signal was sent and acknowledged in two minutes. By September, twenty-four more stations linked the Admiralty in London with the Dockyard in Portsmouth. Signals to and from the Western Squadron in Plymouth were passed along the chain of coast signal stations (see page 39) to Portsmouth and thence by telegraph to London.

Code sheet for Gamble's 'Radiated Telegraph' from his 1797 essay 'Different Modes of Communication by Signals'. Each diagram had space for an updated code, allowing an infinite variation, provide everyone had the current version.

In the same year, Richard Edgeworth came back again with a single-pole telegraph but there is no evidence it was ever used. More spectacular was the seven-arm gantry structure, anticipating by some fifty years the first railway signals. It was demonstrated at the Tuillerie Gardens in Paris but again there is no evidence that its 823,543 possible permutations found favour over Chappe's system.

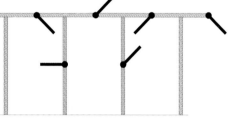

Seven-arm semaphore demonstrated at the Tuillerie Gardens in Paris, 1796. With each arm able to show six positions, or be hidden behind the structure, the number of possible combinations was 7^7.

1800 M. **Charles Depillon,** a former artillery officer, submits proposal for a three and four-arm coastal telegraph. It is the first device to be described as a *sémaphore* but the network he envisaged for the French Atlantic coast was not approved until a year after his death in 1805. Nevertheless the 'Depillon' semaphore outlasted the Chappe system by nearly a century, the last one closing in 1937.[10]

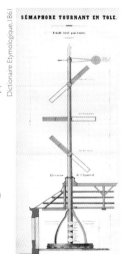

SÉMAPHORE TOURNANT EN TOLE.

Diagram of Depillon semaphore from the *Dictionaire* of 1861 shows a 50ft (15.3m) steel column with an additional disque d'orientation that increased the number of signal permutations from 301 to 2,401. The whole structure rotates on roller bearings at the base.

1806 The Murray telegraph is extended to Plymouth with the addition of a further twenty-two stations, opening on 4th July and allowing transmission and acknowledgement of a time signal from the Admiralty in three minutes.

| 1807 | Sir Charles Pasley, a captain in the Royal Engineers, proposes his first 'Polygrammatic' (meaning capable of signalling more than one letter or number at a time) Telegraph. |

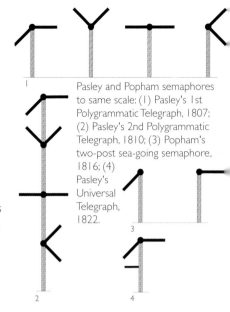

Pasley and Popham semaphores to same scale: (1) Pasley's 1st Polygrammatic Telegraph, 1807; (2) Pasley's 2nd Polygrammatic Telegraph, 1810; (3) Popham's two-post sea-going semaphore, 1816; (4) Pasley's Universal Telegraph, 1822.

| 1808 | Colonel John McDonald publishes his *Treatise on Telegraphic Communication* (see page 14), later arguing that any system '...that cannot express any three figures from 1 to 999 simultaneously is absurd'.[11] Stressing the advantage of a coded system, he advocates a thirteen-shutter telegraph, dismissing Depillon's French semaphore as 'a wretched contrivance'. |

| 1810 | Pasley introduces his 2nd Polygrammatic Telegraph, having had the opportunity of studying the French single-pole system as firsthand during the Walcheren Expedition of the previous year on which he published an account in *Tilloch's Philosophical Magazine*. |

| 1814 | With Napoleon held captive on the island of Elba, Murray's shutter telegraph, always intended as a temporary measure, and the network of coastal stations are closed down, only to be promptly re-instated the following year with the Emperor's escape from Elba. Within eleven days of the Battle of Waterloo, plans are drawn up for a permanent installation. |

Popham's single-pole semaphore replaced Murray's shutters on the roof of the Admiralty in Charing Cross (present-day Whitehall) in 1816.

| 1816 | Rear Admiral Sir Home Popham's single-pole, two-arm semaphore, probably based on the Depillon masts, early versions of which had been captured from the French, is adopted by the Admiralty with a reduced-height two-pole version supplied for use at sea. |

| 1817 | John Macdonald publishes details of what he calls the 'British Semaphoric Telegraph' comprising three double arms on a single 60ft (18.3m) mast with the angles controlled by lines to the end of each arm rather than by pulley round a central pivot. In the same year he proposes his 'Minor Semaphoric Telegraph' that has no less than six arms on a single pivot capable of 2,592 'distinct and simple combinations'. In describing this and |

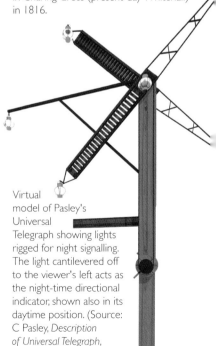

Virtual model of Pasley's Universal Telegraph showing lights rigged for night signalling. The light cantilevered off to the viewer's left acts as the night-time directional indicator, shown also in its daytime position. (Source: C Pasley, *Description of Universal Telegraph*, (London, T. Egerton, 1823)

Macdonald's other inventions, H P Mead could not resist the conclusion that '...it was as well for the sanity of all concerned that the scheme never matured'.[12]

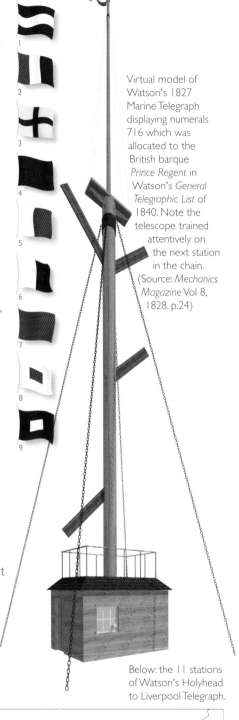

1820 The semaphore line to Chatham is made permanent using Popham's semaphore.

1822 Semaphore established between London and Portsmouth with new locations and structures replacing the Murray shutter telegraph.

In the same year Charles Pasley, now a lieutenant colonel, publishes details of his 'Universal Telegraph'. It is the forerunner of the eventual seagoing masthead-mounted semaphores with trials carried out on ships at the Nore and on the mast of an East Indiaman, the *Vansittart*.

1826 Construction begins on the new Admiralty semaphore to Plymouth but is discontinued after five years with news of developments of the electric telegraph.

1827 The first commercial telegraph is inaugurated between Holyhead and Liverpool using a three-arm single-pole semaphore and telegraphic code developed by a former lieutenant of the Oxford Militia, **Bernard Watson**. To allow communication with ships at sea, Watson provided, for a fee, his own set of numerical flags (see right) with which merchant ships could pass information to their owners through any of the semaphore stations.

The Admiralty, possibly as the result of some partisan lobbying, opts for Pasley's semaphore in place of the Popham system on HM ships.

Virtual model of Watson's 1827 Marine Telegraph displaying numerals 716 which was allocated to the British barque *Prince Regent* in Watson's *General Telegraphic List* of 1840. Note the telescope trained attentively on the next station in the chain. (Source: *Mechanics Magazine* Vol 8, 1828, p.24)

Below: the 11 stations of Watson's Holyhead to Liverpool Telegraph.

842 First railway semaphore introduced on the London to Croydon Railway, approved by the new Inspector General of Railways, none other than Major General Sir Charles Pasley.

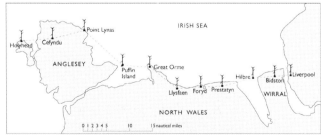

Getting the Message

The arrival of the electric telegraph In 1847 brought about the closure of Popham's two-arm semaphore between London and Portsmouth. Watson's Holyhead to Liverpool Marine Telegraph and, from 1839, that between Spurn Head and Hull survived until 1861. By then the single wooden mast had been replaced by two shorter latticed iron masts with Lieutenant Watson's own code flags replaced by Marryat's code in 1845.

The timeline on the previous pages covers a relatively short period: a little over thirty years from the inauguration of Chappe's telegraph in France to the adoption by the Royal Navy of Pasley's single deck-mounted column for use at sea. Of all the systems trialled in Britain, it was Sir Thomas Pasley who paved the way for the alphabetic semaphore still in use today.

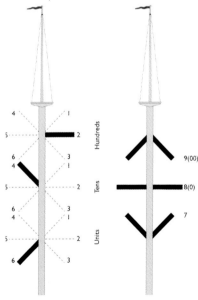

The three arms of Watson's semaphore could signal any number from 1-999; numbers beyond were indicated by class followed by three digits.

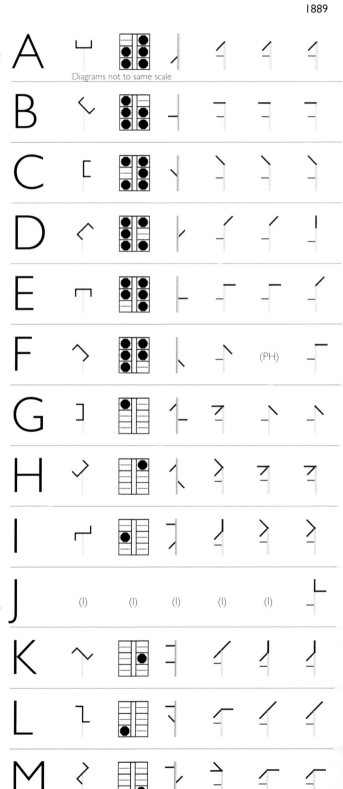

Diagrams not to same scale

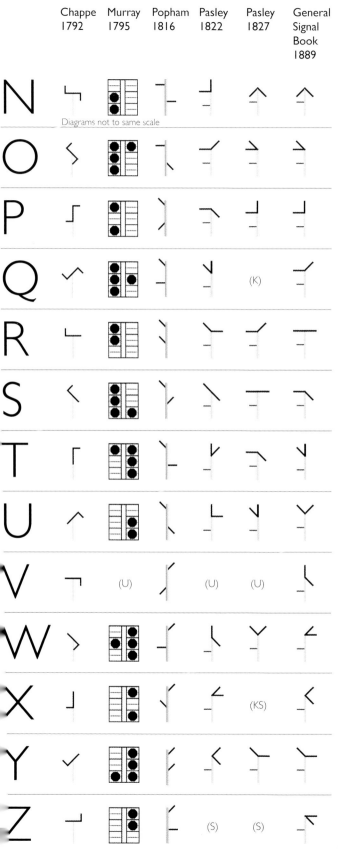

	Chappe 1792	Murray 1795	Popham 1816	Pasley 1822	Pasley 1827	General Signal Book 1889
N		Diagrams not to same scale				
O						
P						
Q					(K)	
R						
S						
T						
U						
V		(U)		(U)	(U)	
W						
X					(KS)	
Y						
Z				(S)	(S)	

Although Pasley had himself, with his early polygrammatic telegraphs, proposed complex permutations of numerical codes, for which he was more than happy to advocate the use of Popham's *Vocabulary Signalling Book*,[1] by 1822 he had abandoned that approach in favour of an alphabetic system. The two arms of his Universal Telegraph offered 28 positions of which five: 17, 35, 26, 4 and 67 (see diagram below right) were reserved for governing signs indicating spelling, numerals or indexed vocabulary. When adopted for use at sea in 1827, three further governing positions were added to align the semaphore more closely with the *Vocabulary Signalling Book* and the remaining 20 positions adjusted with more letters doubled up. The chart on these pages reveal thirteen shapes of the eventual full alphabet were in use at that date, the remaining letters allocated when semaphore instructions were transferred to the *General Signal Book* in 1889.

Below: semaphore alphabet from 1939 edition of Nichol's *Seamanship*.[2]

THE INTERNATIONAL CODE OF SIGNALS   641

SEMAPHORE ALPHABET.

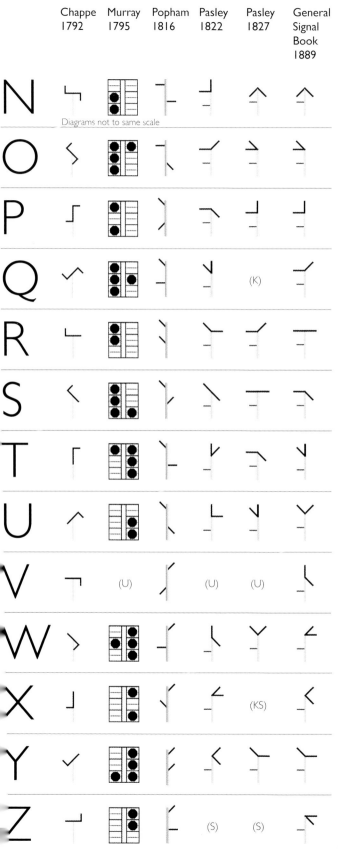

67

Semaphore Goes to Sea: Signals looking for a purpose

Four-flag hoist followed by Naval Code E (International Code J) indicating 'I am going to communicate with you by semaphore'. The signal letters (GQMJ in the 1915 Naval Code) are those of the Dartmouth training ship HMS *Britannia*.

Virtual model of how the three-arm version of Pasley's semaphore with its reflective light panel might have looked, with night view (inset right). (Source: Sketch after *Illustrated London News* in Mead Part V, p.38)

The introduction of Sir Charles Pasley's seagoing semaphore enjoyed some initial enthusiasm, perhaps in response to the frustration some had found in Rear Admiral Popham's twin posts, about which Pasley had himself been particularly critical. Instructions for its use appeared unchanged through five revisions to the *Vocabulary Signalling Book* but from Commander Mead's 1935 account,[1] there appeared to be no consensus as to what it was for: communication with fixed stations ashore, station keeping, manoeuvring signals or ship-to-ship messages spelled out in plain language?

A full half-century after its introduction, the Admiralty invited opinion on the most suitable sizes for the semaphores which were only then beginning to gain wider support. Two sizes were chosen and issued from the dockyards in June 1876: a 15ft (4.6m) mechanically operated version and a much smaller portable version with the arms worked by hand as in the instruction to cadets aboard HMS *Britannia* at the turn of the century (above right). In 1888

some liberties were taken with Pasley's design and a third arm added to extend the number of governing signs when used for signalling manoeuvres, together with a reflective illuminated panel at the rear of the machine to enable night working. It was a short-lived experiment; by 1897, three-arm machines along with their electric light were declared obsolete and semaphore no longer used for manoeuvring. With the Admiralty carrying out simultaneous experimentation in Morse code and coloured light signalling, covered in the following section, it seems in retrospect a bizarre path to have followed, when semaphore had already proved its worth in daylight ashore and afloat.

Right: HMS *Nile* c.1892 showing two semaphores on her foremast.

Sir George Tryon's flagship HMS *Victoria* in dry-dock in Malta, 1892. Note the bridge wing semaphores.

From 1898 new ships were being fitted with a standard two-arm mechanical semaphore on a 15ft (4.6m) column. These were normally fitted on both bridge wings (see HMS *Victoria* above), though sometimes mounted at the aft conning position, and could be trained to continually face the receiving station. The indicator arm was gradually phased out and this type of semaphore was still in service during the Second World War.

Despite experiments with mast mounted semaphores as early as 1822, it would be another sixty-five years before further trials were carried out with two pairs of semaphore arms fitted to the mizzen masts of two elderly armoured frigates *Minotaur* and *Agincourt*. Within a further two years, three-arm semaphores were fitted to battleships and some others, either on existing masts as with HMS *Nile* (see left), or on specially fitted signalling masts. These arms, two pairs of which were fitted, one facing fore and aft and the other athwartships, were operated by chains and sprockets within the mast

and with minimal counterbalance on the 6ft (1.8m) arms must have been a considerable challenge for the signalmen. With three arms, each with a cross piece near the end to distinguish it from the mast in the upright (4) position, these semaphores were primarily used for manoeuvring signals in daylight, though two arms could be used for alphabetic messages. Mead suggests that the move towards mast-mounted semaphores with enclosed mechanisms was made in anticipation of war and the need to replace flag signalling with something more robust.[2]

Whether or not encouraged by the three-arm experiments, it was not until 1895 that the first truck-mounted semaphores were fitted. Often acknowledged as the invention of Sir Arthur Wilson, at that time a Rear Admiral in command of the experimental torpedo squadron, this two-arm semaphore comprised steel arms 12ft (3.6m) long by 15in (.38m), again operated by an even longer system of chains and pulleys within the mast which resulted in what was descibed as a 'considerable backlash' between the operating levers and the semaphore arms. Given the step between the lower mast and the topmast on HMS *Minerva* (see below) this is not hard to imagine, nor is the physical effort required which, in harbour, was exercised daily between 0715 and 0745.

Signalmen aboard HMAS *Akuna* using two-arm mechanical semaphore c. February 1940.

Vice Admiral, later Admiral of the Fleet, Sir Arthur Wilson. In 1897, he succeeded Sir John 'Jackie' Fisher as Third Naval Lord and Controller of the Navy.

Below: The protected cruiser HMS *Minerva* exercising her masthead semaphore in harbour while serving with the Channel Squadron in 1899. She is making the signal 'N' (46) used by scouting cruisers to address units closer to the battle fleet.

With rapid developments in wireless telegraphy from 1902 onwards, the days of the masthead semaphore for long-distance signalling were numbered and within a decade they were phased out completely, leaving the more manageable bridge-wing semaphores. But they were not always available or could not be brought to bear, so signalmen improvised either with their arms or hand-held flags. From the mid-1880s these could be variety of coloured patterns before International Code flag 'O', introduced in the 1901 code, was adopted as the most suitable for signalling against any background. For the first time in a hundred years, arms were again doing what machines had been contrived to imitate and they continue to do so to the present day.

Below: Hand semaphore exercise aboard Training Ship HMS *Britannia*, c. 1903, from a set of postcards published by Gale and Polden.

Bottom left: Cover of US Navy Signal Manual, 1982.

Bottom right: One of fifty cigarette cards in Wills's 'Signalling' series, 1911.

Wig-Wag at War

Brigadier General Albert J Myer c. 1880

By the time hand flags had been adopted for semaphore signalling in the Royal Navy, signalling with hand flags was well established on the other side of the Atlantic from the early days of the Civil War, with the adoption by both the US Army and Navy of a system developed by Albert J Myer. They became known as Wig-Wag Flags.

While semaphore required two flags, Myer employed one flag only with a sharp swing to the left denoting '1' and to the right, '2'. A vertical movement up and down in front of the signalman denoted '3'. The alphabet was encoded with permutations of '1's and '2's with one, two or three '3's denoting end of word, sentence or message.

Flags were issued to signalmen in three sizes: 2ft, 4ft and 6ft with colour variations suited to different conditions.[1] The chosen flag was tied to a hickory staff supplied in 4ft jointed lengths; the combination preferred by the newly formed Signal Corps being the 4ft white flag on a 12ft pole. The red flag was more generally used at sea, with the smaller flags reserved for covert signalling.

SIGNALMAN 3 & 2

NAVAL EDUCATION AND TRAINING COMMAND
RATE TRAINING MANUAL AND NONRESIDENT CAREER COUR
NAVEDTRA 10135-E

WILLS'S CIGARETTES.

G

Critical Information: Semaphore exchanges at Jutland

Although semaphore exchanges made up only 6% of the signal traffic during the 24 hours of 31st May 1916, the record allows tantalising eavesdrops into conversations that would be inhibited by the code book or inappropriate for wider broadcast. With the exception of a general signal at 0400 from Vice Admiral Jerrams in HMS *King George V* on the northern wing of the Grand Fleet, warning of the presence of submarines, almost all are addressed to individual ships. The abstracts that follow are taken verbatim from the official Reports of Proceedings[1] and reveal a mix of the mundane, the dramatic and the poignant. In each case the time of origin, the sender and recipient are given with a note of context.

0428 SO BCF to *Tiger* and *Barham*: Tiger to repeat all signals from Admiral to 5th BS

This is a crucial instruction which bore on later events – see pp 48/49.

0813 *Benbow* to *Bellerophon*: Our Gyro Compass has gone wrong. Our steering I'm afraid is a bit erratic.

HMS *Benbow* was in the van of the 4th Division of the Grand Fleet with *Bellerophon* immediately astern.

0916 SO BCF to *Turbulent*: What did you see when you reported submarine? *Reply:* Very distinct periscope steering south on port quarter.

HMS *Turbulent* was one of 12 destroyers forming the 13th Destroyer Flotilla attached to Beatty's Battle Cruiser Fleet. She was later sunk in a brave night action with the battleship HMS *Westfalen*.

1433 *Indomitable* to SO 3rd BCS: Have just heard Telefunken [German W/T] signals very loud.

The 3rd Battle Cruiser Squadron under Rear Sir Horace Admiral Hood was attached to, but scouting ahead of the main battle fleet.

1446 SO BCF to *Champion*: Send two destroyers to *Engadine*.

The light cruiser *Champion* was leading the 13th Destroyer Flotilla 1.5 miles ahead of the Battle Cruiser Fleet. This signal was immediately followed by a signal by lamp to *Engadine* 6.2 miles ESE ordering her to send up seaplanes to scout to the northeast.

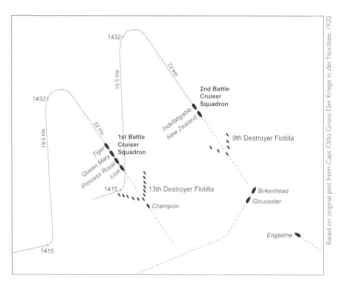

Part of plot for 1415-1445 GMT from High Seas Fleet official records shows disposition of 13th Destroyer Flotilla ahead of 1st BCS and *Engadine* to the south east. See also plot and narrative on page 48.

1530 *Onslow* to *Engadine*: Onslow and *Nestor* [13th Destroyer Flotilla] have been told off to stand by you. Can you please give me any idea of what you are likely to do? *Reply:* Am keeping close to Battle Cruisers. Course N.N.E. Seaplane has gone out E.N.E. and am keeping lookout for return. If you are detailed to work with

71

me please open out to one mile on either side and when picking up planes, should like you to keep at least a mile off. Circle round at 10 knots. My present speed is 20 knots.

HMS *Engadine* was a former cross-Channel steamer converted to a seaplane carrier. This exchange reveals the raw novelty of working with seaplanes, one of which had just made the first ever enemy sighting from a fixed-wing aircraft, reporting the position of Hipper's 1st Scouting Group. After the war she returned to her peace-time service and was later sold to new owners in the US. She was sunk by a mine off Manila in December 1941.

HMS *Engadine* at anchor c.1915. Note the Short Brothers 184 seaplane on her after deck.

IWM SP413 [PD-UKGov]

1612 *Onslow* to *Engadine*: Can you dispense with my services? If so I will join 5th BS. *Reply*: Yes, certainly.

This brief exchange belies the mounting tension of the moment with Beatty's battle cruisers heavily engaged and soon after the 5th BS had opened fire.

1645 SO BCF to *Princess Royal*: Report enemy's battlefleet to C-in-C bearing S.E.

By this time, Beatty's flagship HMS *Lion* had lost her main antenna and relied on semaphore to pass this crucial message via *Princess Royal*, directly astern. Her repeat of the signal by W/T was immediate and gave position, though was logged in Jellicoe's flagship as '...26-30

battleships... bearing S.S.E., steering S.E.'

1722 *Tiger* to SO 1st BCS [*Princess Royal*]: Aft 6" magazine flooded, two guns out of action.

This routine damage report, followed up with a later semaphore exchange reporting difficulty steering, conceals the devastating effect of gunfire on Beatty's battle cruisers with both *Indefatigable* and *Queen Mary* sunk within the first forty minutes of the action.

1750 SO 1st BS to C-in-C: Gun flashes and heavy firing on starboard bow.

The Senior Officer 1st Battle Squadron was Vice Admiral Sir Cecil Burney, Jellicoe's second-in-command and flying his flag in HMS *Marlborough* on the southern wing of the Grand Fleet which was yet to deploy on the port column in line ahead.

Vice Admiral Burney's flagship HMS *Marlborough*. During the course of the action she was badly damaged by a torpedo from the German cruiser SMS *Weisebaden*, forcing Burney to move his flag to HMS *Revenge*.

1800 SO 1st BS to 5th Division [*Collosus, Collingwood, St Vincent* and *Neptune* in the column immediately on his port side]: Remember traditions of Glorious 1st June and avenge Belgium.

This was a reference both to *Marlborough*'s battle honours in Admiral Lord Howe's famous victory on 1st June 1794 and to neutral Belgium to whom Britain was bound by treaty to defend against German agression in 1914.

1802 SO 1st LCS [*Galatea*] to SO 3rd LCS [*Falmouth*]: I was told to keep in touch with the Battle Cruisers. It seems to be getting a bit thick this end. What had we better do?

The 1st and 3rd Light Cruiser squadrons were attached to and scouting ahead of Beatty's battle cruiser Fleet. It was *Galatea* that made the first sighting report of Hipper's battle cruisers. HMS *Falmouth*'s reply is not recorded.

HMS *Galatea*, photographed in 1914

1900 *Canterbury* to SO 3rd LCS: May I join up with you? Reply: Yes.

The cruisers *Canterbury* and *Chester* were accompanying the 3rd Battle Cruiser Squadron attached to Jellicoe's Grand Fleet with the intention that they should join with Beatty's battle cruisers when the action began. By 1900 *Chester* had fallen in with Jellicoe's 2nd Cruiser Squadron leaving *Canterbury* exposed to the east of the fleet, hence the request to join up with the 3rd LCS.

1920 *Warrior* to *Engadine*: Have a bad list and cannot stop engines.

HMS *Warrior* [1st Cruiser Squadron] had been badly damaged in a failed attack against overwhelming odds in which Rear Admiral Arbuthnot's flagship *Defence* was sunk.

2000 *Warrior* to *Engadine*: We are nearly stopped. Come and take in tow.

Engadine succeeded in passing a tow at 2040 and for twelve hours made progress at 7 knots to the west, before taking off the 743 surviving crew members, with *Warrior*'s decks almost awash. She later foundered.

2015 *Princess Royal* to SO BCF: I think *Princess Royal* must have run over a submarine. Ship is not making water. There was a very heavy bump.

A testament to the ever-present fear of submarines although later records show that none were present.

IWM Q 21940 [PD-UKGov]

Armoured cruiser HMS *Warrior* approaching Portsmouth Harbour.

Semaphore in Popular Culture: Protest and public spaces

As with the flags of the International Code, semaphore has also found expression in popular and contemporary visual culture, particularly in the world's most widely appropriated emblem of protest.

Originally described by the designer, Gerald Holtom, a lifelong pacifist and conscientious objector, as the figure of a man despairing at the destructive power of nuclear weapons, the familiar motif embodies the semaphore letters N and D for 'nuclear disarmament', the *raison d'être* of the Campaign for Nuclear Disarmament that adopted Holtom's design before the first Aldermaston march in 1958.

More recently, a broader celebration of the inventions of Murray, Chappe and Edgeworth can be seen in the work of London based Argentinian artist Amalia Pica whose work 'Semaphores' (right) was shown at London's Kings Cross in 2019. Another artist and print-maker exploring the visual language of signalling and information design is James Brown, one of whose prints combining semaphore, morse and phonetic alphabets is shown here.

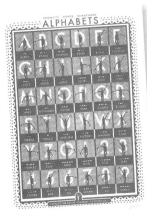

Light Signalling

66 ...it would still be wrong in principle for the Officer, who is responsible for the ship and the lives under his care, to trust the correct signal to a comparatively irresponsible signalman. **99**

1897 memorandum from Lt Alan Everett to the Admiralty Signals Committee advocating a fixed coloured light signalling in place of flashing lights that required a signalman to read and interpret.

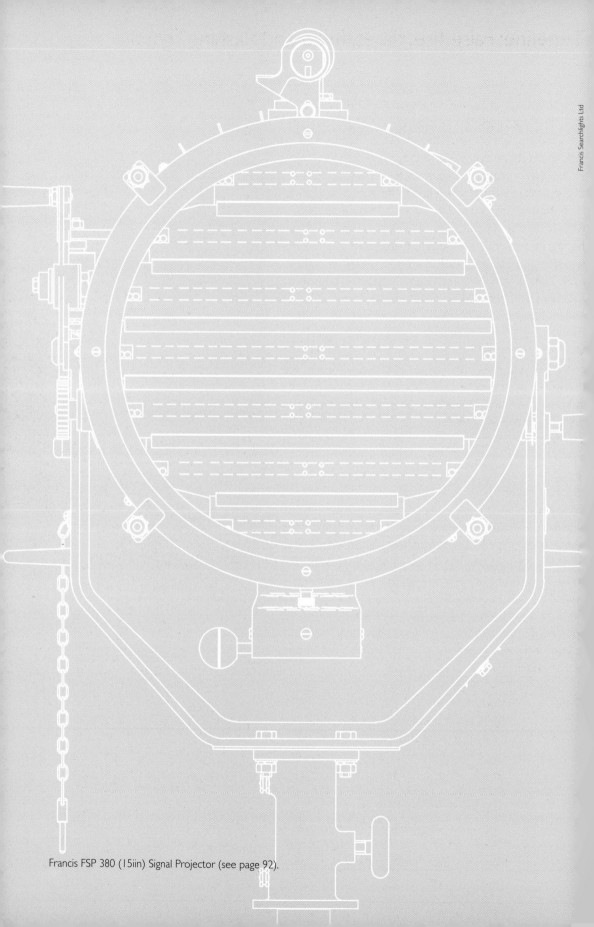

Francis FSP 380 (15iin) Signal Projector (see page 92).

Timeline: False fire, flares, fixed and flashing lights

Light signalling at sea has a timeline that closely maps to that of flag signalling from antiquity and shares many of the same *dramatis personae,* though detail records are sketchy. References to fire signals and flashing sunbeams occur in Homer, Herodotus and Thucydides, and Polybius (c.208-c.125 BC) describes a system attributed to the Trojan mythological hero **Aeneas** that uses two 5x5 matrices of letters. Messages were spelled out by placing the appropriate number of flaming torches along each axis to identify the letter – a system that would only work at night though, if ever used, more likely ashore than at sea.[1] The method does, however, anticipate by almost two millennia the chessboard matrices employed in some 18th-century flag signalling codes.

We know that rockets and flares were used in China to communicate between land armies in the 13th century and that by the early 16th century flashing light signals were included in **Antoine de Conflan**'s *Ordonnances et signes pour naviguer jour et nuyt en une armée royale* (see page 9). The later part of the story is well documented but necessitates a brief diversion into the invention of the electro-magnetic telegraph; for the story of light signalling at sea from the mid 19th century to the present day was driven by the quest for a visual equivalent of the land-based telegraph codes. As with the development of semaphore, the story is populated with a similar mix of scientists, inventive and ambitious naval officers, a Royal Academy-trained American portrait painter and another American inventor and entrepreneur whose proposals finally fell victim to Admiralty dockyard regulations. Without doubt, there were night signals in use at sea in the 16th and 17th centuries even if only for recognition and collision avoidance, but by the mid-18th century we have clearer evidence, which is where this account begins.

1746   John Millan's pocket-sized *Signal Book for the Royal Navy* (see page 11) lists thirty-three manoeuvring signals using combinations of up to four lights hoisted in specified places throughout the rig from the mizzen peak to the bowsprit end. Each signal was to be accompanied by a set number of guns fired '... only from the same side so that the sound may not alter'.[2] Some also called for the burning of 'false fire' – short-lived flares of uncompressed

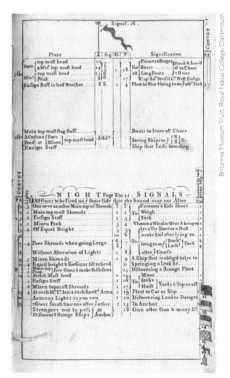

Page from John Millan's pocket signal book shown at just over half actual size. The hand coloured signal flags, lower right, are part of the tabbed index to the different sections of the book.

gunpowder, probably mixed with magnesium powder to enhance their brilliance among the clouds of smoke. The sighting of strange ships, for example, called for one, two, three or four lights vertically where they could best be seen for ships in the NE, NW, SE and SW quadrants respectively. Numbers of ships sighted were to be indicated with corresponding numbers of false fires in rapid succession.

1776 **Admiral Lord Howe**'s first signal book contains several pages of night signals tabulated according to the number of lights employed with configurations given for up to five lights in line, vertically or horizontally, in triangles, squares and rectangles. A separate section lists signals to be made by private ships [with no flag officer embarked] either in response to the flagship or to report sightings or danger. Most signals required the simultaneous firing of a gun or guns in either slow (20 seconds) common (6 seconds) or quick (1.5 seconds) time, though these could be substituted with false fires '...as in the presence of the enemy it may be impractical to make signals by firing a limited number of guns'.[3] Several additions were made

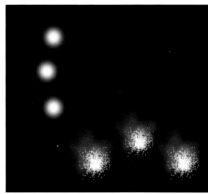

Representation of signal denoting three strange ships sighted to the SE from Millan's pocket book.

Below: Tabulated night signals in Admiral Lord Howe's 1799 *The Signal Book for the Ships of War*.

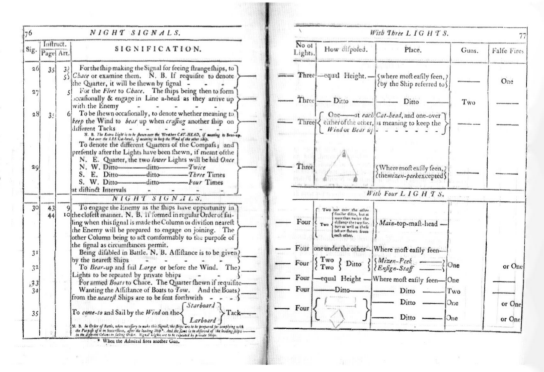

to Lord Howe's signal book in the following decade of almost continuous war at sea and a complete revision published as *The Signal Book for the Ships of War* in 1793, the year before Howe's famous victory of the 1st June. The pages shown at the foot of the previous page are from the 1799 edition of the same book; it contains sixteen pages devoted entirely to night signals.

1777
Alongside a numbered matrix of flag signals **Sir Charles Knowles** includes in his *Set of Signals for a Fleet* a similar numerical system for light signals with false fires used to indicate tens and ordinary lamps for units, a system fraught with risk of mis-reading given, as L.E. Holland points out, the notorious unreliability of false fires.[4]

1782
Rear Admiral Richard Kempenfelt also addresses night signalling in his signal book with a proposal based on a French system. Numbers one to four were to be indicated by lights, five by a rocket and numbers six to nine also by lights doubled up with false fires; tens were indicated by the appropriate number of guns.

1816
The Admiralty's *Vocabulary Signalling Book* based on the work of **Sir Home Popham** included a tabulation of 63 possible combinations of either lights only in various numbers, rockets or false fires, guns alone and guns with rockets or false fires with the most important signals relying on lights and false fires, only, thus avoiding the use of guns.

1837
Building on the pioneering work on electromagnetism of **Hans Christian Øersted** in Denmark and **William Sturgeon** in England, **William Cooke** and **Charles Wheatstone** secure the first patent for an electrical telegraph, demonstrating it on the London to Birmingham Railway. It was the world's first commercial telegraph and used pulses of electrical current to elegantly deflect needles, any two of which would point to a single letter of a reduced alphabet of twenty letters. Later versions of the telegraph employed two needles only, requiring a numerical code.

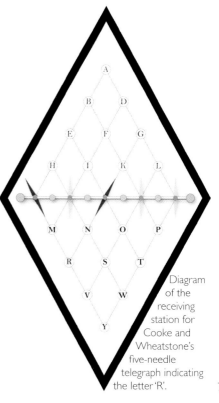

Diagram of the receiving station for Cooke and Wheatstone's five-needle telegraph indicating the letter 'R'.

Samuel F B Morse is granted a patent in the United States, applied for in April 1838, for his American Electro-Magnetic Telegraph.[5] Having studied at the Royal Academy in England in his early twenties, Morse was at this time better known as a portrait painter but a chance meeting with a Boston scientist **Charles Thomas Jackson** on a voyage back from France in 1832 set him on a path that led to the invention of his telegraph and the code that bears his name. It was probably the death of his first wife while he was working on a commission away from home, news of which did not reach him until after she had been buried, that drove his search for a faster method of communication. But it would be a further four years before he was able to secure funds to demonstrate his telegraph to members of Congress. Helping him to build the trial link between Baltimore and Washington, was Baltimore engineer **Henry J Rogers**, whose 'American Calm and Storm Signals' are discussed on page 22 and whose earlier *Telegraphic Dictionary and Seaman's Signal Book* was endorsed in a warm testimonial by Samuel Morse, now elevated to Superintendent of Electrical Magnetic Telegraphs for the United State, in July 1845.[6]

Samuel F B Morse with his receiver in 1857.

Chart comparing three iterations of Morse's code. The Morse/Vial original used variable intervals in some characters (C, O, R, Y and Z) with dashes double the length of dots and the overall length of each letter signal in inverse proportion to its frequency in the English language. This chart, based on Vial's own drawing now in the Smithsonian Institute Archives (SIA 2011-0828), is often misrepresented with several characters, notable B and V reversed. Friedrich Gerke's 1848 revision, which became known as the 'Hamburg Code', increased the ratio between dot and dash to three with a constant interval between each element of the signal. The major change in the final version adopted by the ITU in 1865, and still in use today, is in the coding of the numerals.

Detail from original US patent filed in 1840 shows an early version of alphabet coding.

Morse's original intention had been to transmit numbers only, using a vocabulary code book [possibly Rogers's?]. It was a collaboration with **Alfred Vail**, the son of a New Jersey ironmaster, who also helped fund the work, that persuaded him to include an alphabetic code. Both letters and numbers are included in the 1840 patent (above) and it was Vail who manned the Baltimore end of the line for the demonstration in May 1844. Unlike Cooke and Wheatstone's telegraph, Morse's transmissions caused a stylus to mark a paper tape by passing an electric current through it; the resulting 'dots' and 'dashes' then de-coded into letters and numerals. Vial's 1840 code was quickly adopted in Europe and in 1848 was refined by **Friedrich Gerke**, who had worked on the Cuxhaven to Hamburg Telegraph – a mirror of Watson's telegraphs at Liverpool and on the Humber. Within twenty years and after some further refinement, Morse Code, as it became known, was adopted as the standard for international communication by the International Telegraph Union (ITU). From this foundation, before its natural adoption for wireless telegraphy, followed the competing proposals on both sides of the Atlantic for a robust representation of the code by light. The ensuing debate between flashing and fixed lights would dominate visual signalling at sea for the remaining years of the century.

1859   Although included among a comprehensive set of proposals put forward by **John McArthur** in 1792 (see page 13), it was not until the 1859 edition of the *Admiralty Signal Book* that coloured lights make an appearance and then only in an appendix with the advice that they were not a substitute for General Night Signals but '...a means of communication when the distance does not exceed two miles'.[7]

In the same year, **Martha Coston**, whose story is told in the following pages, was granted a patent in the United States for her night signalling flares that were adopted in large numbers by the US Navy at the outbreak of the Civil War.

861   American inventor **William Henry Ward**, who had previously patented a bullet-moulding machine, self-publishes in London a second edition of his *Ward's Ocean Marine Telegraph*, in which he proposes vertical arrays of three or four white lights for numerals and alphabet respectively. Each light could be covered by a red sleeve or occulted completely by working a series of pulleys. There is no record on either side of the Atlantic of his system being adopted.

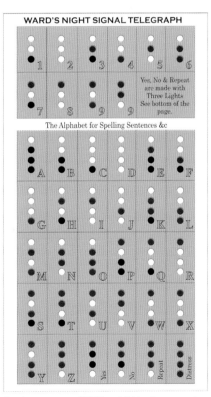

Re-drawn copy of William H Ward's numerals and alphabet from Plate IV of his 1861 *Ocean Marine Telegraph*. The black dots represent the lamp occulted by a shade, the red by a red glass sleeve. Further illustrations show these lamps precariously hoisted to the masthead with similar arrangements on signal stations ashore.

1862 Lieutenant **Philip Colomb** RN, while serving as Flag Lieutenant to Rear Admiral Sir Thomas Pasley (first cousin once removed of Sir Charles Pasley) at Devonport Dockyard, patents the first of his Flashing Signals which he goes on to develop over the next decade in collaboration with Captain **Frank Bolton** of the 12th Regiment of Foot. Colomb and Bolton were expert self-publicists and wasted no opportunity to assert how experiment had shown '...that it is the only system of Flashing Signals which is fully efficient'. [8] They go on to patent several variants that improved the brilliance of the light but it was the keyboard controlled shutter mechanism that enabled reliable and consistent signalling at night that won them the praise from Sir James Anderson laying the transatlantic cable aboard the SS *Great Eastern* in 1866 and Sir Robert Napier during the 1868 Abyssinian campaign.

Besides his collaboration with Francis Bolton on their flashing lights, Lieutenant Colomb was also busy on a study of the tactical possibilities opened up in the transition to steam power that became known as 'steam tactics'. In 1868 he was appointed to the committee set up to revise the *Signal Book* and in 1870, after a spell in command of the sloop HMS *Dryad* on anti-slavery duties in the Arabian sea, he was brought back to the Admiralty to work on a *Manual of Fleet Evolutions* which embodied many of his ideas on Equal Speed Manoeuvres exercised by the Channel Squadron in the previous two years. On its publication in 1874, Andrew Gordon quotes Richard Hough's assessment that it read '... like a glossary of ballet evolutions; elaborate, complex, spectacular and prohibitive of all initiative'. [9] It placed Colomb, by now captain, firmly in the tactical straightjacket that would later involve him in fierce debate with Captain George Tryon when both sat on the Signals Committee in 1878.

1870 Flashing light signals appear for the first time in the Admiralty *Night Signalling Book*, but their use was limited to numerals and procedural

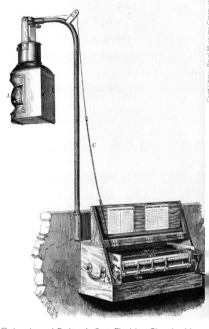

Colomb and Bolton's first Flashing Signal with a shutter activated by pre-set cams on a rotating drum. Later versions used a similar method to control a jet of air from bellows via a rubber tube to blow a small amount of 'Chatham Powder', a mix of magnesium, resin and an inert powder across the flame of a spirit lamp. This gave a light '...second only to lime light in its brilliance' and became known as the Chatham Lamp. With the powder compound available in four strengths, its makers claimed a range of up to twelve miles.

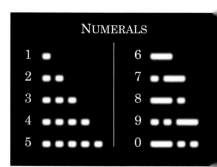

Colomb and Bolton's 'Table of Flashes for all Signal Books' was accompanied by 13 auxiliary or procedural signs denoting Compass, Pennant Number (for addressing specific ships), Horary, Alphabetical and Geographic. (Source: see note 8, p.7)

signs only using Colomb and Bolton's code. Some fixed light signals were included but there was no mention of coloured lights.

1889

Morse Code and procedures for its use by signal lamp at sea is now included in the *General Signal Book* alongside instructions for the electrically illuminated semaphore (see page 68).

In the same year Vice Admiral Sir John Baird, C-in-C Channel Squadron and flying his flag in HMS *Northumberland*, runs trials of a fixed light system for signalling alterations of course at night. Three pairs of red and white incandescent lights (80v, 100 candlepower for the red and 60 candlepower for the white) were fixed on outriggers on the after side of the mizzen mast and controlled in permutations of twos and threes from a plug board on the after bridge to indicate alterations to port or starboard in increments of 4 points [45°].[10] This description is contained in a hand-written submission and detail drawings from *Northumberland's* own artificers to the Admiral Superintendent at Chatham Dockyard and survives among a number of similar submissions of varying practicality in what Andrew Gordon, perhaps unkindly but not inaccurately, termed '...the surge of garden-shed inventions which possessed the Royal Navy in the 1890s'.[11] The essential debate was between fixed pairs of coloured lights and, if so, how many pairs, and flashing lights. Both sides had their champions and detractors.

1891

Following exercises in European waters and the Mediterranean during the previous year, the US Navy Squadron of Evolution under Rear Admiral **John G Walker** experiments with the French-manufactured **Ardois** fixed light signals. The system comprised five pairs of evenly spaced red and white lights suspended vertically, each light controlled from a keyboard. They could be used with a binary code (all white, all red) in the same way as Myer's Wig-Wag code or in any combination either with a pre-concerted code representing, for example,

HMS *Northumberland* photographed in 1890 while flagship of the Channel Squadron.

Reconstruction from original in National Archive of copperplate tabulation of two and three-light signal code submitted from HMS *Northumberland*.

The US Navy Squadron of Evolution, also known as the White Squadron. The flagship USS *Chicago* is in the foreground. Left to right astern are USS *Yorktown*, USS *Boston* and USS *Atlanta*.

compass bearings or with combinations of red and white to represent the dots (red) and dashes (white) of the Morse code. Walker's Flag Lieutenant **Sidney Staunton**, charged with evaluating the Ardois system, reported that '... it fully confirmed the high opinion which I had formed of its capacities... during my cruise in the Mediterranean'.[12] But others disagreed, arguing for a simpler, cheaper and less error-prone four-light system; they also sought a return to Myer's code which had been replaced by Morse in 1886.

1895 'Garden-shed' inventions were by no means the sole province of officers. Writing to the Secretary to the Admiralty from his flagship HMS *Alacrity* at Shanghai in March 1895, Vice Admiral Edmund Freemantle puts forward (with some reservations of his own) a proposal by Petty Officer First Class **Joseph Chandler** for a similar system with four pairs of red and white lights only. The letter encloses Chandler's sketch of the proposal, together with an endorsement of his own commanding officer, demonstrating how it could be used to transmit Morse messages. But an Admiralty note at the foot of the Admiral's letter, before it got as far as the Signals Committee, shows a clear preference for the maintenance of the masthead flashing light for Morse transmission.[13]

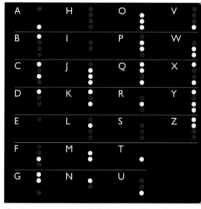

Code proposed by PO Joseph Chandler matching red 'dots' and white 'dashes' to the Morse code.

1897 A long memorandum from Lieutenant **Alan Everett,** later to serve two spells as Captain of the Royal Navy Signals School, to the Signal Committee in Devonport demonstrates the strength of feeling between the two camps. On the one hand, the ease of working with a fixed light system '...so simple that anyone, merely by using his common sense, can immediately read the signal',[14] against the transient nature of Morse by flashing light which required a trained signalman to read and interpret. The nub of Everett's argument comes in the quotation at the start of this section – to what extent was it right to place the burden of correctly reading a signal to execute a critical manoeuvre on the shoulders of a 'comparatively irresponsible signalman'?[15] Setting aside the precarious indicators of social class this reveals, there may be some merit in the argument, not least in his reasoning over the saving in time in making and acknowledging a signal from the flagship. This point is reiterated less than a fortnight later when Vice Admiral Compton Domville, President of the Signals Committee, forwards a recommendation to the Admiralty endorsing the spirit and intent of Everett's memorandum.[16]

Beginning in February of the same year, 1897, lengthy exchanges of letters and telegrams with an American inventor **Mr C V Broughton** of Buffalo, New York, show that the door wasn't yet closed to fixed-light systems. Broughton's proposal was similar to the Ardois four-light system; where it differed was in the keyboard, patented as the 'Broughton Telephotos' (right), and in the signal code employed. The Admiralty agreed to trials of the Broughton signals on ships of the Channel Fleet and, despite some minor reservations in a letter of December 1897, the C-in-C Vice Admiral Sir Henry Stephenson concludes a long letter from his flagship HMS *Majestic* at Arosa in Spain with the recommendation: '...I would strongly urge that this system be fitted to all battleships.'[17] Broughton is July given an order for four more sets which he gratefully acknowledges in a letter of April 1898, excusing a short delay with the explanation that '...trouble in America [the Spanish American War had just broken out following the sinking of USS *Maine* (right) in Havana Harbour] has caused us to run our factory day and night in order to meet the wants of the United States Navy for these instruments...'[18] The letter advises that a supply of an improved cable is being sent by express so that it can be fully examined and tested. It was but, despite having been successfully tested for several months at sea, it met with a response from the Chief Constructor at Portsmouth Dockyard that every contractor fears: 'This class of cable is considered to be of no use for general purpose in the Service...'.[19] There appears to be no further reference to Broughton's Telephotos system.

By the end of the decade light signalling by Morse code using all-round masthead and directional shutter-controlled searchlights (by night and day) was well established and although W/T, then in its infancy but growing fast, would take on some routine and longer-distance signal traffic, visual signalling by light, flag and semaphore remained the only secure means of direct fleet communication.

National Archives ADM 1/6/64

National Archives ADM 1/6/64

Above: Broughton's alphabet code; letters T and U only are shared with Morse code. Left: The Broughton Telephotos keyboard. Repeater lamps on the inside of the cover confirmed the correct lights were displayed. Below: USS *Maine* in 1898. Note the Broughton signal lights rigged on the starboard side of her foremast.

National Museum of the US Navy Lot 3370-7 [PD-US]

Mrs Coston's Telegraphic Night Signals

When Rear Admiral David D Porter manoeuvred his ships off the Confederate-held Fort Fisher in the early hours of 13th January 1865, the night signals he relied on where those developed by a remarkable woman, Martha J Coston.

By the outbreak of the American Civil War in April 1861, flag signalling in the US Navy was well established and, with the addition of Myer's 'Wig-Wag' system, coded visual communications by daylight were routine. By night, however, before the development of reliable shutter systems for signalling by electric light, the centuries-old practice of lantern signalling by permutations of colour, position and number was challenging in all but the calmest conditions. With the transition from sail to steam, the corresponding increase in speed and the precision needed in manoeuvring fleets unfettered by contrary winds, the need for a swift and reliable means of night signalling became paramount. It was the pursuit of this to which Martha Coston would devote much of her life.

Born Martha J Hunt in Baltimore in 1829, Martha moved with her family to Philadelphia where at the age of 16 she

Rear Admiral David D Porter's squadron bombarding Fort Fisher, 13th to 15th January 1865. He later wrote to Coston praising the success of her night signals.

Martha Coston from a full length portrait c.1880. Artist and exact date unknown.

met and fell in love with a young Boston scientist, Benjamin F Coston, who was in charge of the Naval Pyrotechnics Laboratory at Washington Navy Yard. Sources differ on the date of her marriage to Benjamin, but by the time he died in November 1848, probably from excessive exposure to toxic gasses inhaled during his work, his widow, still only 21, had four children to care for. Tragically, two of them died along with her mother within two years.

In straightened circumstances and, by her own admission in her autobiography[1] 'her own ignorance and the duplicity of others', she began a search of her husband's papers, amongst which were proposals for pyrotechnic night signals. Discovering that some prototypes had been made and left in the care of a naval officer, Coston eventually secured their return and persuaded the Secretary of the Navy Isaac Toucey to have them tested. The tests were not a success, but, with encouragement from Toucey who could see 'that the invention... would be of incalculable service to the Government'[2] she set about trying to realise the potential of her late husband's invention. Without any knowledge of chemistry or of business practice, she was totally reliant on the practical assistance and guidance of others – all, of course, male.

In time she had perfected a 'pure white and a vivid red light'; she still needed a third colour, preferably a patriotic blue, to enable Benjamin's proposed three-colour signal code to work. The turning point came while watching the New York firework celebrations marking the completion of the transatlantic cable in 1858. Posing as a man for fear that her requests would not be taken seriously, she wrote to several of the pyrotechnic manufacturers who had given the display to try and find the blue she was after, or failing that, a 'brilliant green'. Within ten days she had received a reply with a sample of a green compound that proved successful and wasted no time in establishing a partnership with the New York firm of G A Lilliendahl, later to become the Coston Signal Company, to manufacture the composite cartridges. Twelve of these in different colour combinations, together with a hand-held device from which they could be safely fired were the essential components of the Coston Telegraphic Night Signal System.

Under the direction of Secretary Toucey, a Navy examination board rigorously tested the signals in all their different permutations and endorsed them as

'decidedly superior'. They had passed the test and orders for multiple sets of flares immediately followed, though the Navy hesitated to buy the patent outright, which became the subject of protracted debates in Congress. On 5th April 1859, Martha Coston was granted a patent which, probably for sound business reasons, given her husband's established reputation as an inventor at the Navy Yard, she registered in her husband's name.[3] Subsequent refinements, the first in 1871 (detail left), were registered in her own name and further developments of the Coston signal system were registered by her two surviving sons Henry H and William F Coston in 1877 and 1901.[4]

With Congress prevaricating over the purchase of her patent, Martha, having secured patents on her invention in several European countries, set off for Europe where she stayed for two years negotiating the sale of patents to the British and French governments,

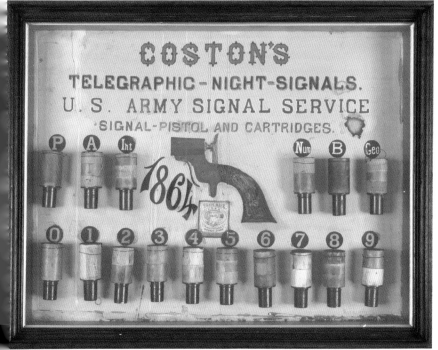

A display case of Coston's Telegraphic Night Signals and Signal Pistol adapted for use by the US Army. The cartridges were fitted into the pistol and fired with a percussion cap and remained in place until the colour combinations had burned through before inserting the next flare. The set normally comprised ten numeral flares, a preparatory signal and an answering signal; this set has four additional signal flares. Though dated 1864, the display is signed at bottom right by Martha's son 'W F Coston, [18]76'

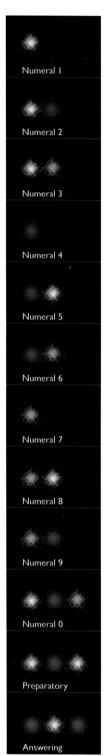

Numeral 1

Numeral 2

Numeral 3

Numeral 4

Numeral 5

Numeral 6

Numeral 7

Numeral 8

Numeral 9

Numeral 0

Preparatory

Answering

Sequence of colours representing each numeral in the Coston Code of 1859-1864. Each colour lasted approximately 8 secs.

the latter successfully. By the time she returned home in June 1861, the Civil War was two months old and she re-opened her petition to Congress for the sale of her patent which, she argued, would prove 'a valuable auxiliary for the Navy'. By August a bill authorising the purchase was passed; the begrudging price of $20,000 (approximately $500,000 in 2018)[5] was exactly half what she had asked for. By then her company had already supplied flares to 600 ships of the Union Navy.

Fort Fisher fell on 15th January 1865, closing off the Confederates' last access to the sea at Wilmington. Admiral Porter was later to write to Martha Coston singing the praises of what must by then be regarded as *her* signals and the crucial role they had played in the deployment of his fleet. Praise for the Coston signals also came from commanders of blockading fleets off Louisiana, along the Mississippi and the Atlantic seaboard.[6] But the greatest testimony to Martha's dogged determination to build on her husband's work is revealed in the latest iteration of the Coston flare – a tricolour signal projectile adapted for use in a standard US Army grenade launcher (below) – for which a patent was registered in March 2014.[7]

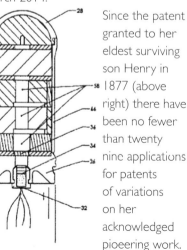

Since the patent granted to her eldest surviving son Henry in 1877 (above right) there have been no fewer than twenty nine applications for patents of variations on her acknowledged pioneering work.

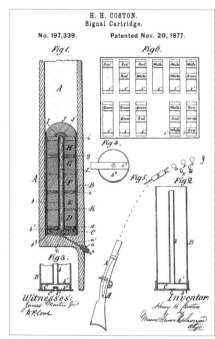

Diagram accompanying Henry Coston's patent application of November 1877, for a Coston flare to be fired from a gun, releasing its coloured lights in the correct sequence at a given height. This, the specification rightly claims, 'possesses great advantage over the stationary signals because the point signalled from cannot be discovered by the enemy.' It is no surprise that this proposal was quickly taken up by the United States Life Saving Service that later became the US Coastguard Service.

Below: 1913 press advertisement for Coston's Marine Signals, the opening line of which asserts 'The only signal recognised by the British Board of Trade...'.

What Ship, Where Bound? Procedures and protocols

All forms of visual signalling had their own procedures to ensure signals were correctly addressed, understood and, when required, specific actions executed. Light signalling was no exception and although the lead in the development of signalling by flashing light was taken by the Royal Navy, it was readily adopted by merchant ships, with competition among shipping companies for the most proficient.

An Admiralty Fleet order of 1936 calls upon H.M. ships, while acknowledging 'the difficulties under which the merchant navy labour [with limited bridge personnel]' to '...lose no opportunity of carrying out signalling exercises with British merchant ships'.[1] There was even an Admiralty form S.174 (Quarterly Return of Signalling with British Merchant Ships) in which to record visual signalling traffic. A 'remarks' column encouraged comments on speed and accuracy and a quarterly abstract released to the Press which, the AFO goes on to assert, 'has undoubtedly stimulated keenness in signalling efficiency throughout the Merchant Navy'.[2]

Merchant ships were required to carry an all-round masthead signal light, many carried Aldis lamps and some also had 10in signal projectors on the bridge

Wartime production at the Bolton factory of Francis Searchlights Ltd.

wings. Warships were rather better equipped with an array of pedestal-mounted and hand-held signal lamps. They could also deploy a team of signalmen under the Chief Yeoman on permanent standby, their work perhaps familiar from the distant flashing signals seen in wartime newsreel footage. Among their signalling arsenal were 20in carbon arc searchlights, capable of daylight signalling at extreme range, 15in and 10in incandescent projectors, the 5in hand-held Aldis lamp (shown in use, right), used for both day and night signalling and smaller lamps down to the 1in Hether binocular-mounted signal lamp. It was used for night signalling when the fleet was darkened and could be fitted with blue or red filters to reduce light levels to an absolute minimum.[3]

Diagram of Hether lantern, storage box and battery case from instruction manual.

While many signal exchanges in the merchant service were initiated by warships, the majority were routine exchanges between passing ships, often useful sources of weather and sea-state ahead; some seasonal exchanges; some idle curiosity. Nearly all began with the familiar question from which this book takes its title, but not all followed the expected pattern, as the exchange with a US warship on the follow pages shows.

Page of procedural signals from 1961 edition of Brown's Signalling.

Shore-based Wrens signalling with an Aldis lamp, 1944.

FSP 127 Mark V Aldis lamp still in production today.

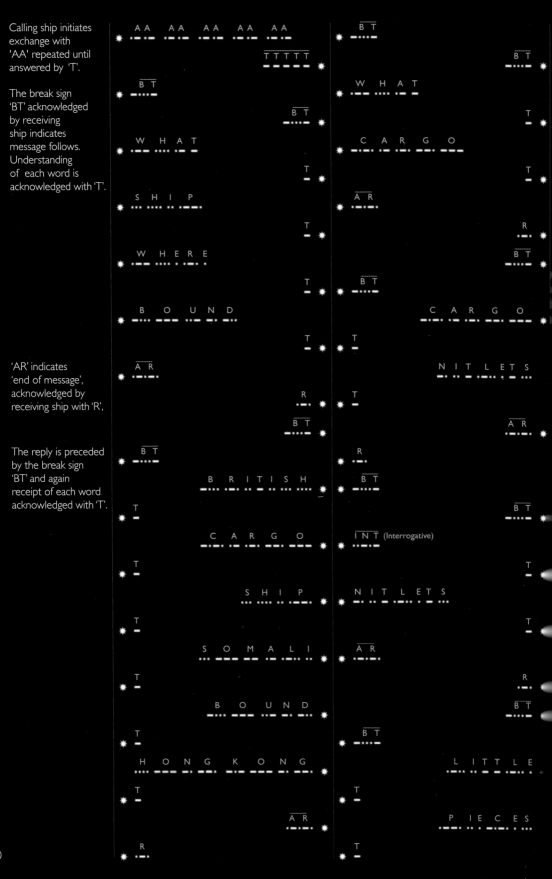

Calling ship initiates exchange with 'AA' repeated until answered by 'T'.

The break sign 'BT' acknowledged by receiving ship indicates message follows. Understanding of each word is acknowledged with 'T'.

'AR' indicates 'end of message', acknowledged by receiving ship with 'R',

The reply is preceded by the break sign 'BT' and again receipt of each word acknowledged with 'T'.

A A   A A   A A   A A   A A
T T T T T
B T
B T
W H A T
W H A T
C A R G O
S H I P
A R
W H E R E
R
B T
B O U N D
C A R G O
T   T
A R   N I T L E T S
R   T
B T   A R
B T   R
B R I T I S H   B T
T   B T
C A R G O   I N T (Interrogative)
T   T
S H I P   N I T L E T S
T   T
S O M A L I   A R
T   R
B O U N D   B T
T   B T
H O N G K O N G   L I T T L E
T   T
A R   P I E C E S
R   T

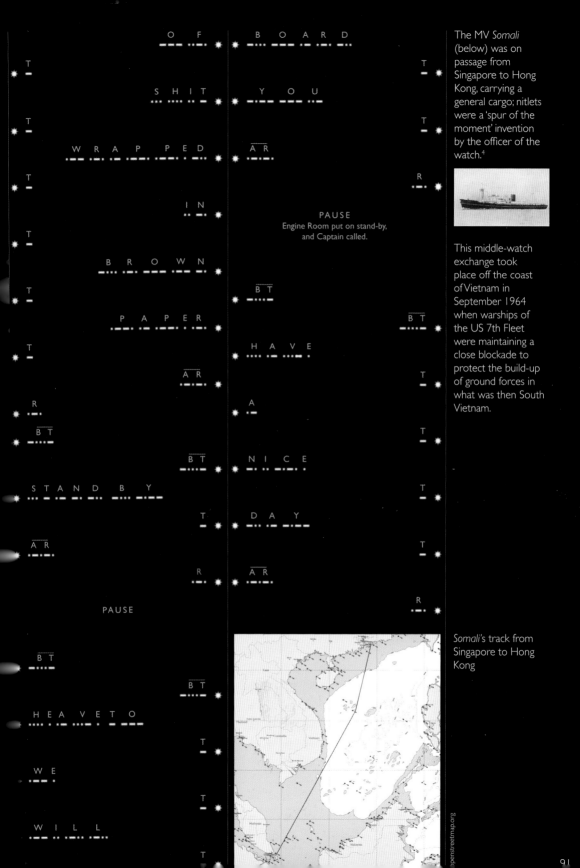

O F   B O A R D

T   T

S H I T   Y O U

T   T

W R A P P E D   A R

T   R

I N

T   **PAUSE**
           Engine Room put on stand-by,
           and Captain called.

B R O W N

T   B T

P A P E R   B T

T   H A V E

A R   T

R   A

B T   T

B T   B T N I C E

STAND BY   T

T   D A Y

A R   T

PAUSE   A R

           R

B T

B T

HEAVETO

T

W E

T

W I L L

T

The MV *Somali* (below) was on passage from Singapore to Hong Kong, carrying a general cargo; nitlets were a 'spur of the moment' invention by the officer of the watch.[4]

This middle-watch exchange took place off the coast of Vietnam in September 1964 when warships of the US 7th Fleet were maintaining a close blockade to protect the build-up of ground forces in what was then South Vietnam.

Somali's track from Singapore to Hong Kong

Francis Searchlights Ltd

One of four 15in (380mm) signal projectors installed on the Royal Navy's new aircraft carrier HMS *Queen Elizabeth*.

Far right: 15in signal projector in use aboard HMS *Ark Royal* during Operation 'Telic', the invasion of Iraq, in 2003.

Right: Experimenting with FLTC aboard the destroyer USS *Stout* at Naval Station Norfolk.

An exchange like this with its formal procedural structure would today be carried out by VHF radio or by e-mail via satellite uplinks. But visual signalling by light, in both the visible and invisible (infra-red) spectrum, though used less frequently, still has a role to play. Just as with the US Navy's Squadron of Evolution trials of fixed-light signalling and parallel trials carried out by the Royal Navy at the end of the 19th century, so current experimentation in the United States is adapting visual signalling practice to maintain its relevance in the 21st century – particularly in reducing a ship's tell-tale electronic signature.

Rising to the challenge to develop an intuitive interface that would allow text messaging by signal lamp between ships, the US Navy's Office of Naval Research began trialing a Flashing Light to Text Converter (FLTC) in 2017. Using a GoPro style camera to capture Morse code messages, a tablet-based algorithm decodes and converts the signal to text as on a smartphone. Replies are tapped on the screen and the process reversed with the software automating the signal by lamp. The result is quick and error-free and, while still using Morse code nearly 180 years after it was first patented, no longer requires a trained operator.

At the time of writing (March 2020) a second phase of trials is under way replacing the shutter and xenon lamp with LED arrays capable of much faster modulation of intensity with corresponding increase in speed and capacity. Speeds of up to 1,200 words per minute are theoretically possible.[5]

We may learn that, more than 500 years since Antoine de Conflans first advocated flashing lights for signalling at sea, visual signalling has a long future.

Ike and the Inspector: Morse code in popular culture

Unlike signal flags and semaphore, light signalling by Morse code does not lend itself so readily to visual interpretation. There is no playful colour of the flags or the tactile exuberance of Amalia Pica's installations, but nevertheless representations of Morse code have also found their way into popular culture and merchandise.

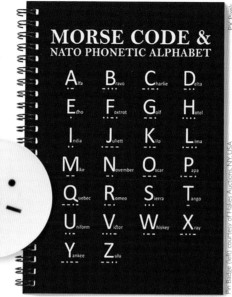

Probably the earliest example comes from the 1952 election campaign contested by General Dwight D Eisenhower, best known for the slogan 'I Like Ike!', in this pin badge shortened to the three characters IKE in Morse code. More recent is Barrington Pheloung's iconic 1987 score for the television series *Inspector Morse* with the subtly coded motif M-O-R-S-E. running through it.

With greater visual impact is the interpretation of the W/T distress signal from the stricken RMS *Titanic* in the Titanic Belfast exhibit depicting the sinking, in which the now familiar S-O-S call was made alongside the original distress call C-Q-D. Rather less dramatic is the usual collection of mugs, tee-shirts

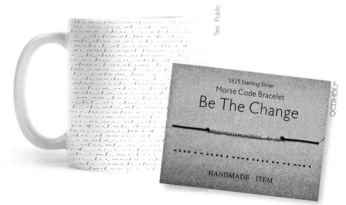

and merchandised messaging that draw with varying degrees of success on a visual representation of Morse code.

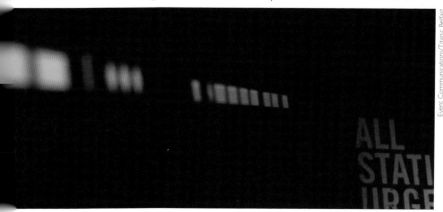

Distress signals in Morse code add dramatic effect to depiction of the sinking of the *Titanic* at Titanic Belfast. 93

Notes

Flag Signalling

Introduction
1   T Ditcham, *A Home on the Rolling Main* (Barnsley, Seaforth Publishing, 2013) p.89

Long Distance Information
1   L E Holland RN 'The Development of Signalling in the Royal Navy', *The Mariner's Mirror*, 1953, 39:1, 5-26
2   Holland, *ibid.*
3   W G Perrin, *British Flags: Their Early History and Development at Sea* (C U Press, 1922) p.146
4   See David Howarth, *Trafalgar, The Nelson Touch* (London, World Books, 1970) p.104,

Calm before the Storm
1   Henry J Rogers *American Code of Signals* (Baltimore, 1854) p.vii, Caird Library.

Ensigns and Etiquette
1   Perrin, p. 40
2   *Ibid.* p. 47
3   *Ibid.* p. 55
4   See Hendrick Vroom *The Return of Prince Charles from Spain 5th October 1623*, Royal Museums Greenwich.,
5   See see detail of Reinier Nooms, *The Battle of Leghorn 1653*, Rijksmuseum, Amsterdam on page 6
6   Perrin, p. 68.
7   *Ibid.* p. 87, referencing State Papers of Henry VIII ccv, 160

Nelson Confides...
1   Howarth, p.154
2   Nicolas, Sir N (Ed.) *Letters and Dispatches of Lord Nelson* Vol III, quoted in Perrin, p.174
3   Howarth, p.155 and Corbett J S (Ed.), *Fighting Instructions 1530-1816* Publications of Navy Records Society Vol XXIX, 1905
4   K G Baker, 'True Story of Nelson's Famous Signal' in *Navy and Army Illustrated*,

October 16th 1896 (sourced from www.navyhistory.org.au)
5   Jonathan Clements *Admiral Togo: Nelson of the East* (London, Haus Publishing, 2010) p..45
6   Howarth, p.155

Commercial Codes
1   See Cmdr H P Read RN, US Naval Institute Proceedings No 361, Vol 59, pp.370-375.
2   W K Stewart, *Brown's Signalling: How to Learn the International Code of Signals* (Glasgow, Brown Son and Ferguson, 1961)
3   J C Dobbin, Secretary to the Navy, Circular, appended to Henry J Rogers *American Code of Signals* (Baltimore, 1854) Caird Library 627.72(73)
4   *Nautical Magazine and Naval Chronicle for 1840* (London, Simkin Marshal and Co, 1840) p.516
5   Rogers, Plate 5
6   Registrar General of Shipping and Seamen, *Supplement to International Code of Signals, 1873* Caird Library 627.724(100)

British Naval Codes
1   Andrew Gordon, *The Rules of the Game, Jutland and British Naval Command* (London, Penguin Books, 2015) p.599
2   Lt Cdr J K Dempsey RN, *The Evolution of Signalling by Flags at Sea* (https://www.com-msmuseum.co.uk/)
3   Surrey History Centre Archive Ref 1226. See also http://www.holywellhousepublishing.co.uk/Tufnell.html
4   *Admiralty Visual Signalling Instructions, Section II* (London, HMSO, 1944) Caird Library 627.724.725

'Land's End for Orders'
1   *International Code of Signals* American Edition (Washington, Hydrographic Office, 1923)
2   Lloyd's Signal Station, Lizard Peninsula in *Engineering Timelines* (http://www.engineering-timelines.com)
3   *Supplement to International Code of Signals, 1873*
4   Frank Kitchen (1990), 'The Napoleonic War Coast Signal Stations', *The Mariner's Mirror*, 76:4, 337-344
5   Arthur C Wardle (1948), 'Liverpool Merchant Signals and House Flags', *The Mariner's Mirror*, 34:3, 161-168
6   *Ibid.* p.163. For more on the history of the Bidston Signal station go to www.bidstonlighthouse.org.uk
7   William Enfield, *An Essay Towards the History of Liverpool* (London, Joseph Johnson, 1774) p.68
8   Source uncertain, quoted in Wardle, p.162

Unintended Consequences
1   R Harding, *Seapower and Naval Warfare* (London, UCL Press, 1999) p.148
2   J Corbett, *Fighting Instructions 1530-1816* (Project Guttenberg, 2005: Publications of The Navy Records Society Vol. XXIX, 1895) p.111
3   B Wilson, *Empire of the Deep, The Rise and Fall of the British Navy*, (London, Weidenfeld and Nicholson, 2013) p.364
4   See 'Clerk of Eldin, A Statement of Facts', 1934 The *Mariner's Mirror* 20:4, pp.475-495

5   G R Barnes and J H Owen (eds) *The Private Papers of John Earl of Sandwich, First Lord of the Admiralty* (1932-38) quoted in Wilson, p.360
6   The London Gazette Extraordinary, 25th May 1780 in *Universal Magazine of Knowledge and Pleasure* (London John Hinton, 1780 Google Books)
7   See Wilson, pp. 360-361 for more on Admiral Rodney's views on his subordinates.
8   Rear Admiral C. Ekins, *Naval Battles from 1744 to 1814 Critically Reviewed and Illustrated* (London, Baldwin Cradock and Joy, 1824/ Google Books)
9   Captain T White, *Naval Researches: A Candid Enquiry into the Conduct of Admirals Byron, Graves, Hood and Rodney* (London, Whittaker, Treacher and Arnott, 1830/Google Books), pp.34-49)
10 Ekins, quoted by White, p.45
11 White, p. 44
12 *Ibid.* p.45
13 Richard Hough, *Admirals in Collision* (London, White Lion Publishers 1959) quoted in Gordon, p.195
14 Letter to Secretary of the Admiralty 21st November 1891, enclosing Temporary Memorandum 'A': A System for Fleet Manoeuvres With and Without Signals (ADM 116/64 Vol 11)

Reaching the Limits
1   Grand Fleet Battle Orders XXIII Signals in Action, ADM 137/288 quoted in John Brooks, *The Battle of Jutland* (Cambridge, Cambridge Military Histories, 2016)
2   See Gordon, Ch. 9
3   *Battle of Jutland Official Despatches* (London, HMSO, 1920), p.10
4   Gordon, p.140

Flags Still Flying
1   *Naval Appendix to International Code, 1935* (Admiralty Signal Department) M4343/35 in Caird Library 627.724
2   Vincil T Clark, *Signalman 3 & 2* (NAVEDTRA 10135-E Naval Educaation and Training Command, 1982) Fig 5.2
3   *Communications Instructions Signalling Procedures in the Visual Medium* (Combined Communications-Electronics Board July 2005), p.1-1

Bravo Zulu
1   See Captain Barrie Kent, RN, *Signal! A History of Signalling in the Royal Navy* (Petersfield, Hyden House, 1993)

Under Starter's Orders
1   For an historical perspective on yacht racing and the rules that applied, see Dixon Kemp, *A Manual of Yacht and Boat Sailing* (London, The Field Offices, 1878), p.382. For a modern day equivalent see *Racing Rules of Sailing* (Southampton, International Sailing Federation (ISAF), 2012)

Semaphore

Timeline

1 Charles F Partington, *The Century of Inventions of the Marquis of Worcester from the Original MS, with Historical and Explanatory Notes* (London, John Murray, 1825. Project Guttenberg/Apple Books, 2015)
2 Quoted in *Mechanics Magazine*, Vol 8, 1828 (London Knight and Lacey, 1828), p. 295
3 This story is related in two sources slightly differently: see *Mechanics Magazine* article, *ibid.*, p. 294 and Russel W Burns *Communications: An International history of the formative years* (London, IEE, 2004) p.35
4 *Ibid.* p.39
5 *Ibid.* p.42
6 See Burns, p.52 and for a detailed and near contemporary description of Edgeworth's telegraph and others, see Abraham Rees (Ed.) *Cyclopedia of Arts Sciences and Literature* Vol XXV (London, Longman, 1819, www.archive.org) pp. 214-215
7 Rev. John Gamble, *Essay on the Different Modes of Communication y Signals* (London W. Miller, 1797/Google Books), p. 71
8 *Ibid.*, p.89
9 *Ibid.*, p.90
10 F Cabane, *Charles Depillon (1768-1805): l'inventeur des sémaphores côtiers* (Tremer, Plouzané, 2007)
13
14 H P Mead RN, 'The Story of Semaphore Part III', *The Mariner's Mirror* 20:2 1934, p. 6. Parts I to IV in *The Mariner's Mirror*.
2 *Ibid.* p. 215

Getting the Message

1 See Mead, 'The Story of Semaphore Part IV' *The Mariner's Mirror* 20:3 1934, p.367
2 *Nicholl's Seamanship and Nautical Knowledge* (Glasgow, Brown, Son and Ferguson), known as 'the Manual', was the indispensable *vade mecum* for all sea-going cadets covering everything from the construction of deep tanks to the Collision Regulations.

Semaphore goes to Sea

1 See Mead, 'The Story of Semaphore Part V', *The Mariner's Mirror* 21:1 1935, pp 34-35
2 *Ibid.*, p.41

Wig-Wag at War

1 For more on Myer, Wig-Wag and the early days of the US Signal Corps, see Rebecca Raines, *Getting the Message Through* (Washington, US Army Center for Military History, 1995)

Semaphore at Jutland

1 See Appendix 2 to *Battle of Jutland Official Despatches* (London, HMSO, 1920)

Timeline

1 Christopher H Sterling (ed), *Military Communications from Ancient Times to the 21st Century* (Santa Barbara, ABC-CLIO, 2008)
2 John Millan, *Signals for the Royal Navy and Ships under Convoy* (London, J Millan, 1746)
3 *The Signal Book for The Ships of War, 1799* quoted by L E Holland RN in 'The Development of Signalling in the Royal Navy', *The Mariner's Mirror*, 1953 39:1, 5-26 p.24
4 *Ibid.*
5 US Patent Office, Patent No. 1647, 20th June 1840
6 Henry J Rogers, *Telegraph Directory and Seaman's Signal Book* (Baltimore, F Lucas Jr., 1845)
7 Holland, p.25
8 P Colomb and F Bolton, *Flashing Signals Adopted in the Navy and Army* (London, Mitchell and Co., 1869), p. 3
9 Richard Hough, *Admirals in Collision* quoted in Gordon, *The Rules of the Game*, p. 186
10 Letter and enclosures from Vice Admiral Sir John Baird to Admiral Superintendent, Chatham (undated), 1889 in ADM 116/64 Case 335, Vol 10
11 Gordon, p.307
12 Timothy S Wolters, *Information at Sea: Shipboard Command and Control in the US Navy from Mobile Bay to Okinawa* (Baltimore, John Hopkins U.P., 2013)
13 Note appended to letter from C-in-C China to Secretary to the Admiralty 12.1.1892 in ADM 116/64

14 Memorandum: *Coloured Lights for Signalling Purposes* from Lt Alan Everett to Signals Committee, Devonport 6.3.1897 in ADM 116/64
15 *Ibid.*
16 Letter from Vice Admiral Compton Domville, President of the Signals Committee to Secretary to the Admiralty 19.3.1897 in ADM 116/64
17 Letter from Vice Admiral Sir Henry Stephenson to Secretary to the Admiralty 9.12.1897 in ADM 116/64
18 Letter to the Secretary to the Admiralty from C V Broughton, 18.4.1898
19 Memorandum from Chief Constructor, Portsmouth Dockyard to Captain John Durnford, HMS *Vernon* 15.9.1898. HMS *Vernon* was the Royal Navy Torpedo School and at that time also had oversight of new signalling equipment and W/T training

Mrs Coston's Telegraphic Night Signals

1 Martha J Coston, *A Signal Success: The Life and Travels of Mrs Martha J Coston* (Philadelphia, Lippincott, 1886)
2 Coston, p.43
3 US Patent Office, Patent No. 23,536, April 5th 1859
4 US Patent Office, Patent Nos. 197,339 and 674,400, 20th November 1877 and 21st May 1901 respectively.
5 Purchasing power calculator, www.measuringworth.com
6 Coston, pp.100 and 101

7 US Patent Office, Patent No. 8,677,904 March 25th 2014

What Ship, Where Bound?

1 Admiralty Fleet Order Volume 1937, Section 1, p. 148
2 *Ibid.* p.149
3 I am indebted to former Royal Navy Signalman David Morris and to the Francis Searchlight Company for their helpful advice on historic and present day signalling equipment
4 A record of this exchange was kindly provided by Commander Nick Messinger RNR, then 4th Officer aboard P&O's MV *Somali*, on which the author also sailed as a cadet
5 Office of Naval Research in *US Navy Institute News* [https://news.usni.org/2017/07/19]

Bibliography and Further Reading

John **Brooks**, *The Battle of Jutland* (Cambridge, Cambridge Military Histories, 2016)

Jack **Broome**, *Make a Signal* (London, Putnam, 1956) and *Make another Signal* (London, Harper Collins, 1973)

Charles H **Brown**, *Nicholl's Seamanship and Nautical Knowledge* (Glasgow, Brown Son and Ferguson, 18th Edition 1938 reprinted 1949)

Russell W **Burns**, *Communications: An International history of the formative years* (London, Institute of Electrical Engineers, 2004)

Jonathan **Clements**, *Admiral Tojo: Nelson of the East* (London. Haus Publishing, 2010)

P **Colomb** and F **Bolton**, *Flashing Signals Adopted in the Navy and Army* (London, Mitchell and Co., 1869)

Sir Julian **Corbett**, *Fighting Instructions 1530-1816* (Project Guttenberg, 2005: Publications of The Navy Records Society Vol. XXIX, 1895)

Martha J **Coston**, *A Signal Success: The Life and Travels of Mrs Martha J. Coston* (Philadelphia, Lippincott, 1886)

Tony **Ditcham**, *Life on the Rolling Main*, (Barnsley, Seaforth Publishing, 2013)

William **Enfield**, *An Essay Towards the History of Liverpool* (London, Joseph Johnson, 1774)

Andrew **Gordon**, *The Rules of the Game, Jutland and British Naval Command* (London, Penguin Books, 2015)

R **Harding**, *Seapower and Naval Warfare* (London, UCL Press, 1999)

Richard **Hough**, *Admirals in Collision* (London, White Lion Publishers, 1959)

— *The Great War at Sea* (Oxford, Oxford University Press, 1983)

David **Howarth**, *Trafalgar, The Nelson Touch* (London, World Books, 1970)

Dixon **Kemp**, *A Manual of Yacht and Boat Sailing* (London, The Field Offices, 1878)

Captain Barrie **Kent**, RN, *Signal! A History of Signalling in the Royal Navy* (Petersfield, Hyden House, 1993)

John **Millan** *Signals for the Royal Navy and Ships under Convoy* (London, J. Millan, 1746)

W G **Perrin**, *British Flags: Their Early History and Development at Sea* (Cambridge University Press, 1922)

Rebecca **Raines**, *Getting the Message Through* (Washington, US Army Center for Military History, 1995)

Henry J **Rogers**, *Telegraph Directory and Seaman's Signal Book* (Baltimore, F. Lucas Jr., 1845)

Nigel **Steel** and Peter **Hart**, *Jutland 1916, Death in Grey Wastes* (London, Cassell, 2003)

Christopher H **Sterling** (ed), *Military Communications from Ancient Times to the 21st Century* (Santa Barbara, ABC-CLIO, 2008)

W K **Stewart**, *Brown's Signalling: How to Learn the International Code of Visual and Sound Signals* (Glasgow, Brown Son and Ferguson, 1961)

Ben **Wilson**, *Empire of the Deep, The Rise and Fall of the British Navy* (London, Weidenfeld and Nicholson, 2013)

Timothy **Wilson**, *Flags at Sea* (NMM/ Chatham Publishing, 1986)

Timothy S **Wolters**, *Information at Sea: Shipboard Command and Control in the US Navy from Mobile Bay to Okinawa* (Baltimore, John Hopkins University Press, 2013)

On-line Resources

www.flaginstitute.org
UK charity working to promote interest in all aspects of flags and flag flying.

www.hampshireflag.co.uk
UK flag maker. Catalogue includes all Naval and International Code flags.

www.seaflags.us
A good source for everything connected with flags used at sea in the United States.

www.crwflags.com
US flag maker and retailer. Links to sponsored site Flags of the World (FOTW) resource and forum for serious vexillologists

www.commsmuseum.co.uk
Royal Navy Communications Branch Museum and Library.

www.distantwriting.co.uk
Private site tracing history of telegraph companies in the UK.

www.nmrn.org.uk
National Museum of the Royal Navy has a comprehensive on-line resource on all aspects of the naval history including flags and signalling.

www.civilwarsignals.org
An extensive archive on all methods of signalling in use during the American Civil War including Martha Coston's Telegraphic Night Signals. For more on Martha Coston see C Kay Larson, 'A Woman with Flare', *New York Times*, 2.11.2012